Brain-Computer Interface

Edited by Vahid Asadpour

Published in London, United Kingdom

IntechOpen

Supporting open minds since 2005

Brain-Computer Interface
http://dx.doi.org/10.5772/intechopen.94618
Edited by Vahid Asadpour

Contributors
Erick Giovani Sperandio Nascimento, Ilan Figueirêdo, Lilian Lefol Nani Guarieiro, Vahid Asadpour, Reza Fazel-Rezai, Mohammad-Reza Akbarzadeh-Totonchi, Maryam Vatankhah, Manupati Hari Nithin Hara Nithin Reddy, Jen A. Markovics, Ibrahim Kaya, Horst Schneider, Sigrid Seuss, Jennifer Riederle, Beatriz García-Martínez, Antonio Fernández-Caballero, Arturo Martínez-Rodrigo, Amitabh Dube, Kapil Gupta, Jitendra Gupta, Bhoopendra Patel, Shubha Dube, Umesh Kumar, Sanjay Kumar Singhal, Kavita Yadav, Lubaina Jetaji, Ebru Sayilgan, Yalcin Isler, Yilmaz Kemal Yuce

First published in London, United Kingdom, 2022 by IntechOpen
IntechOpen is the global imprint of INTECHOPEN LIMITED, registered in England and Wales, registration number: 11086078, 5 Princes Gate Court, London, SW7 2QJ, United Kingdom
Printed in Croatia

British Library Cataloguing-in-Publication Data
A catalogue record for this book is available from the British Library

Additional hard and PDF copies can be obtained from orders@intechopen.com

Brain-Computer Interface
Edited by Vahid Asadpour
p. cm.

This title is part of the Artificial Intelligence Book Series, Volume 9
Topic: Computational Neuroscience
Series Editor: Andries Engelbrecht
Topic Editor: Magnus Johnsson

Print ISBN 978-1-83962-522-0
Online ISBN 978-1-83962-529-9
eBook (PDF) ISBN 978-1-83962-530-5
ISSN 2633-1403

For EU product safety concerns:
IN TECH d.o.o., Prolaz Marije Krucifikse Kozulić 3, 51000 Rijeka, Croatia, info@intechopen.com or visit our website at intechopen.com.

IntechOpen Book Series
Artificial Intelligence
Volume 9

Aims and Scope of the Series

Artificial Intelligence (AI) is a rapidly developing multidisciplinary research area that aims to solve increasingly complex problems. In today's highly integrated world, AI promises to become a robust and powerful means for obtaining solutions to previously unsolvable problems. This Series is intended for researchers and students alike interested in this fascinating field and its many applications.

Meet the Series Editor

Andries Engelbrecht received the Masters and PhD degrees in Computer Science from the University of Stellenbosch, South Africa, in 1994 and 1999 respectively. He is currently appointed as the Voigt Chair in Data Science in the Department of Industrial Engineering, with a joint appointment as Professor in the Computer Science Division, Stellenbosch University. Prior to his appointment at Stellenbosch University, he has been at the University of Pretoria, Department of Computer Science (1998-2018), where he was appointed as South Africa Research Chair in Artifical Intelligence (2007-2018), the head of the Department of Computer Science (2008-2017), and Director of the Institute for Big Data and Data Science (2017-2018). In addition to a number of research articles, he has written two books, Computational Intelligence: An Introduction and Fundamentals of Computational Swarm Intelligence.

Meet the Volume Editor

Vahid Asadpour, MS, Ph.D., is currently with the Department of Research and Evaluation, Kaiser Permanente Southern California. He has both an MS and Ph.D. in Biomedical Engineering. He was previously a research scientist at the University of California Los Angeles (UCLA) and visiting professor and researcher at the University of North Dakota. He is currently working in artificial intelligence and its applications in medical signal processing. In addition, he is using digital signal processing in medical imaging and speech processing. Dr. Asadpour has developed brain-computer interfacing algorithms and has published books, book chapters, and several journal and conference papers in this field and other areas of intelligent signal processing. He has also designed medical devices, including a laser Doppler monitoring system.

Contents

Preface

Nowadays, there is significant scientific and industrial interest in developing a robust interface to connect the human brain to a computer-controlled system. The potential capabilities of such a system include a wide range of disability services, prosthetic organ control, and industrial and military applications. The aim of studies in brain-computer interfacing (BCI) is to provide a working tool for patients with disabilities to communicate neurophysiological activities into physical actions. The human-machine controller expands the degree of freedom and the available options for manipulation and navigation of the system by using direct cognitive commands.

This book presents some recently established methods for processing and deep learning methods for categorizing EEG signals. The chapters cover a range of topics including noninvasive and invasive signal acquisition, signal processing methods, deep learning approaches, and implementation of BCI in experimental problems.

The book provides a comprehensive summary of conventional and novel methods for processing brain signals. These methods include some subcategories of deep learning methods to show the contribution of this methodology to BCI. It also provides an overview of the applications of BCI to highlight the growing idea of interfacing minds with machines. I hope this book will inspire academic readers and researchers with new ideas in this area. I would like to acknowledge the chapter authors for their excellent contributions.

Vahid Asadpour
Department of Research and Evaluation,
Kaiser Permanente,
Pasadena, California

Research Scientist (Former),
UCLA,
Los Angeles, California

Section 1

Introduction to Brain-Computer Interface

Chapter 1

Language as the Working Model of Human Mind

Amitabh Dube, Umesh Kumar, Kapil Gupta, Jitendra Gupta, Bhoopendra Patel, Sanjay Kumar Singhal, Kavita Yadav, Lubaina Jetaji and Shubha Dube

Abstract

The Human Mind, functional aspect of *Human Brain*, has been envisaged to be working on the tenets of *Chaos*, *a seeming order within a disorder*, the premise of *Universe*. The armamentarium of *Human Mind* makes use of distributed neuronal networks sub-serving *Sensorial Mechanisms, Mirror Neurone System (MNS)* and *Motor Mechanisms* etching a stochastic trajectory on the virtual phase-space of *Human Mind*, obeying the ethos of *Chaos*. The informational *sensorial mechanisms* recruit *attentional mechanisms* channelising through the window of chaotic neural dynamics onto *MNS* that providing *algorithmic image information flow along virtual phase- space coordinates* concluding onto *motor mechanisms* that generates and mirrors a *stimulus- specific and stimulus-adequate response*. The singularity of self-iterating fractal architectonics of *Event-Related Synchrony (ERS)*, a *Power Spectral Density (PSD)* precept of *electroencephalographic (EEG) time-series* denotes preferential and categorical *inhibition gateway* and an *Event-Related Desynchrony (ERD)* represents event related and locked gateway to stimulatory/excitatory neuronal architectonics leading to stimulus-locked and adequate neural response. The contextual inference in relation to *stochastic phase-space trajectory of self- iterating fractal of Off-Center* α *ERS (Central)-On-Surround* α *ERD-On Surround* θ *ERS* document efficient *neural dynamics of working memory.*, across patterned modulation and flow of the neurally coded information.

Keywords: Human Mind, Chaos, Stochastic Trajectory, Mirror Neurone System, Neural Dynamics, Electroencephalograph (EEG), Event Related Synchronisation (ERS)/Desynchronisation (ERD)

1. Introduction

1.1 The multi-dimensional hierarchy of organisational levels in brain

Brains are characterised by every property that engineers and computer scientists detest and avoid. They are **chaotic, unstable, nonlinear, non-stationary, non-Gaussian, asynchronous, noisy, and unpredictable in fine grain**, yet undeniably they are among the most successful devices that a billion years of evolution has produced. **Brain systems** operate on many levels of organisation, *microscopic, mesoscopic and macroscopic*, each with its scales of time and space. **Dynamics**, the modelling of change, is applicable to every level, from the atomic to the molecular,

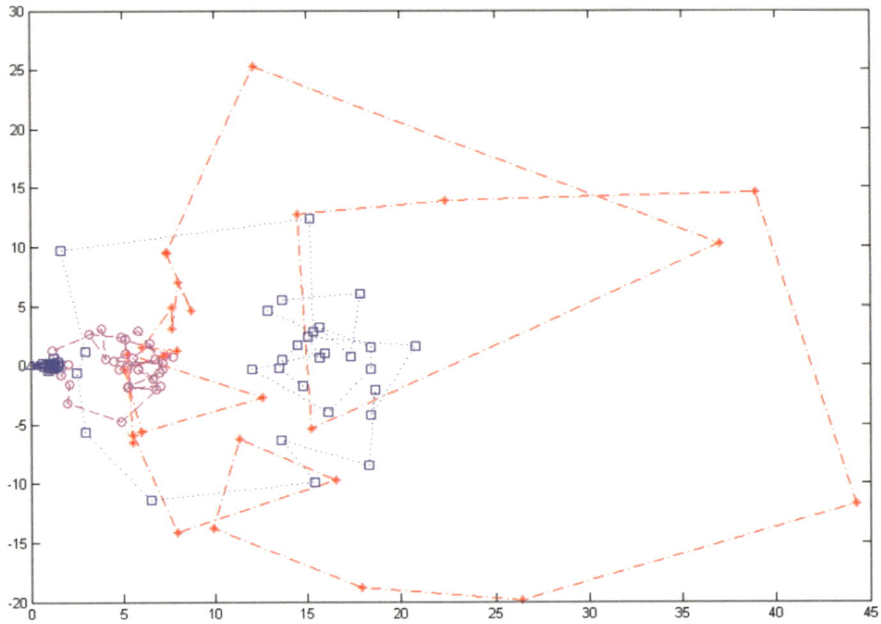

Figure 1.
Hilbert transform of EEG lead pair across EEG lead pair for EEG frequency-waveform bands of δ, θ, α, β, γ representing the stochastic trajectory of neural dynamics in real-time.

and from macromolecular organelles to the neurones into which they are incorporated. In turn the **neurones** form populations, these form the sub-assemblies of brains, and so on up to embodied brains interacting purposively with the material, interpersonal, and politico-social environments.

Subsequently, the ***mesoscopic*** **level**, very aptly characterised by ***nonlinear dynamical electroencephalographic (EEG) electrical activity*** [1], seems to be the optimally suited substratum of interplay of neuronal discharge and its patterning, that seems to have been very beautifully and intelligently decrypted and decoded through the armamentarium of **digital biological signal processing across *linear (relative and absolute power spectral densities, coherence and others)* and *non-linear classifiers (entropy, fractal dimensions and others)*.**

The varied discrete and quantal features of *Human Brain* working and co-opting, in tandem and in sync, across the dimensions and coordinates of space and time evolve into the *phase-space stochastic trajectory of* abstruse and arcane domain of *the Human Mind* observing the principles of *non-linear dynamics of Chaos [2]*. The human brain provides the scaffold and framework for the functional dynamics of human mind in real time [3] following the principles of *Chaos*, further documented by our centre in 2009 [4] (**Figure 1**).

Carl Jung has very aptly outlined the schema *as "In All Chaos There is A Cosmos, In All Disorder A Secret Order"*. The *Secret Order* as has been exemplified by Carl Jung forms the nidus to explore further the realms of *Chaos*.

2. The working of the human brain

The *Human Brain* communicates and interfaces through electrical and chemical processes in a *fractal* and *self-iterating fashion*. The neurones fire at a rate of 5–50/second through *integrate-and-fire neurones and resonate and fire neurones* with a summated thought-processing time of around 329 milliseconds [5].

3. Neurotransmitters

The chemicals deployed by the *Human Brain* involve *neurotransmitters, neurohormones, neuropeptides, neuromodulators inclusive of dopamine, serotonin, acetylcholine, gamma aminobutyric acid (GABA), glutamate, glycine, adenosine triphosphate (ATP)* to name some of the chemicals. The neurotransmitters seem to be the key to functioning and influencing the neurophysiology of the *Human Brain* and are diffusely distributed with selective cerebral predominances responsible for the genesis of a select personality-trait brain waves and rhythms. The precursors to the *neurotransmitters, amino acids*, are readily available in the diet and the *diet* (and its interaction with the specific metabolic patterning of an individual) *determines the persona/qualia* of an individual.

4. The rigid versus distributed functional patterning

- **The Frontal Lobe** has a dominance of **dopamine** and is responsible for the generation of **β**, **beta-rhythm** that represents the state of alertness in an individual,

- **The Prefrontal Lobe** being the site of spiritual experience and consciousness generates the fast brain waves of **γ**, **gamma-rhythm** (with frequency of more than 30/second) through the action of **glutamate** neurotransmitter that excites the fast inhibitory synapses interconnecting inhibitory neurones [6].

- **The Parietal Lobe** has a dominance of **acetylcholine** that engenders the **α**, **alpha-waves**,

- **The Occipital Lobe** with the dominance of the neurotransmitter **serotonin** induces **δ**, **delta-waves** of sleep and memory consolidation and

- **The Temporal Lobe** does the overall function of a tranquilliser and/or analgesic through the help of **GABA** inducing **θ**, **theta wave** pattern along the human mental phase-space.

However, the modular aspect of the *Human Brain* with rigid configurations (as proposed by *Cajal* way back in 1913) has given way to the model of distributed neuronal networks that has resilience and the capacity to adjust and be flexible to the demands of internal and external milieu, wherein the mind-set with positivity influences and modulates the distributed neuronal pools and networks evolving the cognitive abilities of an individual.

A subtle and perceptible paradigm shift has been witnessed across the frontiers of *Neurosciences and Neurology* wherein **the *Human Mind*, once thought to be working along the framework of modular architectonics**, is now envisaged to be traversing the alleyway along the distributed neuronal pools conjuring onto **dedicated and apportioned networks** that have the ability and the interface to **crosstalk**.

The building block scaffold of the respective dedicated neuronal pools is the archetypal *neurone* that has the endowed potential to respond in a ***space and time coordinate-locked precept of action potential, the espoused all-or-none phenomenon*** that incidentally happens to be the singular canonical principle of functional neurones. The armamentarium of neuronal language evolved through the presence and/or absence of *action potential all-or-none phenomena* along with **differential**

neuronal architectonics processing inclusive of serial, parallel, divergent, convergent, reverberating along with inter-neuronal reverberations [7].

The unitary and singular neuronal tenet got segregated through the remarkable neurophysiological characteristic of ***learning*** **into dedicated neuronal pools** that became functionally conspicuous and perceptible as ***sensory, mirror, motor and interneurones***. Such dedicated neuronal pools then evolved the distinctive patterned waveforms as evinced through electroencephalographic (EEG) signals [8] of *theta (**θ**), delta (**δ**), alpha (**α**), beta (**β**) and gamma (**α**) waves* and **such distributed neuronal pools** then evolved discrete neurodynamical phenomena of

- *Event-Related Desynchrony (ERD) [evinced as decrease in Power Spectral Density (PSD)]* and

- *Event-Related Synchrony (ERS) [an increase in PSD]* in respective wave-forms bands.

5. The human mind

The *Human Mind* is the neurophysiological precept that tends to amalgamate the ***Triune Brain Complex*** through the distributed electro-chemical neural circuitry that follow the non-linear chaotic neural dynamics simulating the principles of Chaos in Nature [9]. ***The primacy and singularity of chaos and chaotic systems (Complex Dynamic Systems)*** *depict* ***behaviours of determinism, paradox, self – generation, self – iteration, self – organisation, intrinsic unpredictability within the confines of the defined geometry across space – time*** *that is sustained by the complex feedback loops. The qualia of chaotic systems include the* ***sensitivity to initial conditions*** *with* ***disproportionate responsiveness to stimuli****, the translatability from micro-through mesoscopic and macroscopic proportions, and the* ***attractor-centring that is shuffled across space – time*** *and is apparently a – causal (enfolded; implicate/explicate), global singularity and is flexible and amenable to creation.* ***The Strange Attractor-Centred Stochastic Trajectory*** *so evolved through the* ***neuronal oscillations*** [4, 10] that sublimes the awe and grandeur of *Human Mind* seems to be the gateway and/or portal to flow of information that is legible, reproducible and stands the vagaries and vicissitudes of the flow of space and time.

In this backdrop and the chance brush and close encounter with *Chaotic Nonlinear Neural Dynamics of Human Mind* [4], our centre came across the novel finding of ***Dysfunctional Mirror Neurone System*** *['Broken Mirrors' of Professor V. S. Ramachandran and Oberman [11]]* in children with *Attention Deficit Hyperactivity Disorder (ADHD)*, a disorder of social intelligence, an antecedent sequel to '*Broken Mirrors*', that was neurodynamically represented as the phenomenon of *Event-Related Synchrony (ERS) of mu rhythm (alpha waveform along somatosensory EEG lead pairs)* [12]) when the ADHD participant children aped and imitated the action protocol of hyperventilation, while an *Event-Related Desynchrony (ERD)* was observed in the similar rhythm of *mu waveform* in EEG lead pairs of normal control children.

The *Human Mind* replicates the transmutation and metamorphosis of the non-linear dynamics of chaos wherein a fine interplay between matter and energy takes place, i.e., the abstruse versus the intangible with quantum shift being appreciated through the perturbations of space–time synthesising *sensory–mirror–motor neurones–cognition* tangible precepts plunging along the ethos and tenor of chaos, journeying to the most fundamental or primal state of consciousness – Chaos, when shift in primal image of self becomes possible through its de-structured nature in entirety. In this qualified state of Chaos, the Human Mind evolves onto a rhythm/pattern that seems to be reverberating with Cosmic Consciousness.

It is conceivable that the sensorial stimulus evinces a characteristic ***event/stimulus-related synchrony (ERS)* of *theta (θ) wave-form*** reflective of an antecedent and incidental entrainment of ***attentional neuronal mechanistic resources*** that seemingly feeds onto and opens the portal of the algorithmic flow of *mirror neurone system arsenal through means of* ***event/stimulus-related desynchrony (ERD) of alpha (α) wave-form*** *that seems to feed onto the* **motor neuronal system** *responding through ERD to effect a cogent, logical and stimulus/event-locked response.* Such a model of intricate dance of ***event/stimulus-related synchrony (ERS)* of *theta (θ) waveform*** *and* ***event/stimulus-related desynchrony (ERD) of alpha (α) wave-form*** [13, 14] has been hypothesised to be the mainstay of the working *Human Mind*.

The *Human Mind* is conceived as an entity forming the functional singularity of *Human Brain* that evolves through the integration of *quantum mechanics of wave-particle espousing the inter-convertibility of mass into energy waveform and vice versa, the Higgs Boson being the interface and the amalgamating particle.*

A set of neuronal pools, *referred to as fractals with the inherent capability of self-organising and self-iterating,* are recruited to sub-serve a distinct selected function limited by the coordinates of space–time with a time decay of 2–3 seconds recouped and retrieved by another set of neuronal pools observing similar fractal neurodynamical dimensions of synced *ERD* and *ERS*. The set of neuronal pools that evolve during the course of time rhyme and oscillate with a specific wave-pattern that is construed and translated onto the *stochastic phase-space trajectory with the strange attractor* specific for the function being attended to silhouetting and profiling the *Human Mind*.

Taking the analogy further, *Cosmic Consciousness* seem to be the predicate of *mass-energy wave-form interface* as exemplified by the *God particle, Higgs Boson*. The effervescent and evolving Human Mind works on the same principle of Cosmos with a tendency to cohere and sync with the flow of Cosmic Consciousness.

6. The working model of language

The working of *Human Mind along with its functional and morphological correlates* has been an arena that has overwhelmed and beguiled mankind since times immemorial.

The Neurophysiologists and Cognitive Neuroscientists have resorted varied procedures, both non – invasive and invasive, to gain an insight and to reveal the mystics of *working human mind*, wherein *Electroencephalography (EEG)* and *Event Related Potentials (ERPs)* [15–17] provide the desired armamentarium to record underlying neural dynamics of human mind in real – time, through precepts of flow of space and time namely, amplitude and latency, respectively, that are time-locked to specific sensory, motor and/or cognitive modalities of stimuli [18].

EEG and ERPs seem to be the tools with temporal precision but poor spatial localisation for appreciation of underlying neuronal dedicated networks and their dynamics for various higher mental and cognitive functions to identify, isolate and register across space – time, the physical qualia of the stimulus (features detection, the so-called feature-detectors). The neural dynamics of working memory have been envisaged to be funnelled onto the language acquisition processes and the interplay between multiple frequency wave-forms in the cortical neural networks play an elementary deciding role in such an intricately woven process [19–22].

Neurolinguistics, an interdisciplinary domain that draws in inputs from application disciplines of neurosciences, linguistics, cognitive sciences, computers electronics and communications, neuropsychology and neurophysiology, and basic sciences of mathematics and physics, explores the underlying neural mechanisms of human brain and its correlation with the *phenomenon of the means of communication, that is* **Language**.

7. The ontogeny of language: The piggyback ride of working memory

At birth young infants exhibit a universal capacity to detect differences between phonetic contrasts used in world's language [23]. The mother (or father) has to entrain the attentional mechanisms of the child through *Social Gaze* with subsequent motherese (or fatherse or parentese), a form of language that involves lot of changes in pitch, is melodious and repetitive. *Social Gaze or Eye Contact* with the mother forms the essence or pre-requisite of genesis of language, wherein the *vowels* (and that too the *extremes of vowels, i.e., 'a' and 'o' 'u'*) precede *consonants* for the mere fact that lips movements is maximal for vowels and due to the simplified mechanism(s) that underlie the *neurolinguistics* of vowel. The language development or transition of the human mind onto the axes of language has been hypothesised to take place along two neural phases, namely *Phase I* and *Phase II.*

7.1 Phase I (neurodynamical phase)

The neurodynamical phase also known as the ***general open-system*** is uncommitted and open to change and plasticity and is the phase where priming of the human mind takes place. The universal capacity of the human mind is dramatically altered by the language experience starting as early as 6 months for *vowels (a, e, i, o, u)* and by 10 months for *consonants*. The *extremes of vowels, namely 'a', 'o' and 'u'* involve maximal movements of the lips that the child gets enamoured through the mental landscape so formed by the stochastic trajectory initiated in the *Phase-Space of Human Mind* by the system of *Mirror Neurones.*

7.2 Phase II (linguistic phase)

It represents the language specific phase wherein the human mind becomes committed to the specific language that is being acquired and usually starts from end of the first year of life. Neural oscillations across the coordinates of time (brainwaves), within individual neurones or through interactions among neurones, *are rhythmic or repetitive patterns of neural activity of the central nervous system* and *such patterned neural dynamics signify and describes the respective neurophysiological functional characteristics*. The techniques of *Biological Signal Processing (BSP)* have been employed to classify and categorise EEG signals through linear domain of power spectral density (PSD), linear discriminant analysis (LDA) and varied non – linear domains of neural networks.

The concept of human mind in acquisition of language or general learning mechanism(s) contribute to such an evolved mechanism of spoken and written language that imprisons the mechanistic of mirror neurone system (MNS) and *synaptic neuroplasticity*. *MNS* plays a pivotal role and is considered to be an interface between the qualia of sensorium and motor system of the intricately woven *Human Mind*, wherein activation of *Mirror Neurone System* initiates the process of image formation in the virtual phase-space trajectory of human mind so evolved by the baseline reverberating chaotic neural dynamics, a phenomenon learnt and hard-wired through the neurophysiological process of memory.

The neural signature of *Working Memory (WM)*, *the primacy of emergent Human Mind* [24], for *Encoding, Registration and Retrieval of Memory* [25, 26] inputs has been postulated to be served by *three EEG Wave-Forms Complex of Theta* [27, 28], *Alpha and Gamma frequency bands* [10] with a bootstrapping blueprint wherein the ***gamma wave-forms or bursts hitchhike or piggy back rides the theta wave responsible for feature detection*** **along with *alpha-theta wave-form*** *that coincidentally allocates* ***attentional resources*** *onto the evolved dedicated neuronal circuitry that are stimulus-specific* [29, 30]. These frequency oscillations have been observed to **modulate neuronal excitability** by controlling neuronal firing, and could be responsible for **holding of stimulus-specific information in space and time along the coordinates of working memory neuronal pool** [31]. Such a neural synchronisation proposal may provide a solution to underlying mechanism(s) of synthesis and amalgamation of features of an object through coordinated firing patterns that in essence underlie the *feature detector mechanism(s) of neuronal process* [32].

It has been envisioned that,

- ERS in θ, theta frequency waveform is related to encoding and retrieval of episodic or new information, and

- ERD in α, alpha frequency waveform is related to encoding and processing of semantic information [29, 30, 33–37].

Pfurtscheller and Klimesch [38], Pfurtscheller and Aranibar [39] and Pfurtscheller and Lopes da Silva [40] had reported that during visual stimulation **alpha wave-form desynchronises** giving rise to ***ERD*** **over occipital recording sites** whereas **over motor cortex synchrony in form of *ERS*** could be observed. Sauseng et al. [41], [42] put forward the observation of **change in *PSD of alpha wave-form*** **that is observed at the occipital and pre-frontal areas during *top-down processing*** *in a working memory task*, wherein a ***decrease in alpha PSD power at occipital site*** with a **consequential *increased alpha PSD power is observed at prefrontal EEG electrode site*. The *ERD quantum* of alpha frequency wave-form during encoding** in a visual working memory task has been correlated with the memory load [29, 30, 33–36].

However, [43] reported that the processing of working memory of **encoding and retention involves the oscillatory activities along multiple frequency bands of EEG wave forms inclusive of *alpha frequency*** *as well* through local and long-range neural networks proposing the existence of **multiple parallel functional mechanisms of alpha oscillations** [44]. In this context of equivocal representation of alpha oscillations, it would be interesting to examine changes in alpha oscillations pattern that could be sensitive and characteristic to working memory task.

The observation documented from our laboratory of ***theta wave form band synchrony*, known as *Event-Related Synchrony (ERS)* mirroring increased *PSD*,** across distributed range of task relevant areas of brain namely,

- ***The Retention Function*** being primarily centred along select frontal and temporal areas,

- ***The Semantic Manipulation*** *along select frontal, temporal and parietal* with

- ***The Backward Manipulation*** involving frontal, central and temporal areas [*Neuropsychological Trends* in print] during working memory task of

registration, retention and retrieval processing is reflective of dynamical linking, an observation that had been documented by EEG studies of [41, 42, 45] as well, though [46–48] could not appreciate such breakthrough linkage (**Figure 2**).

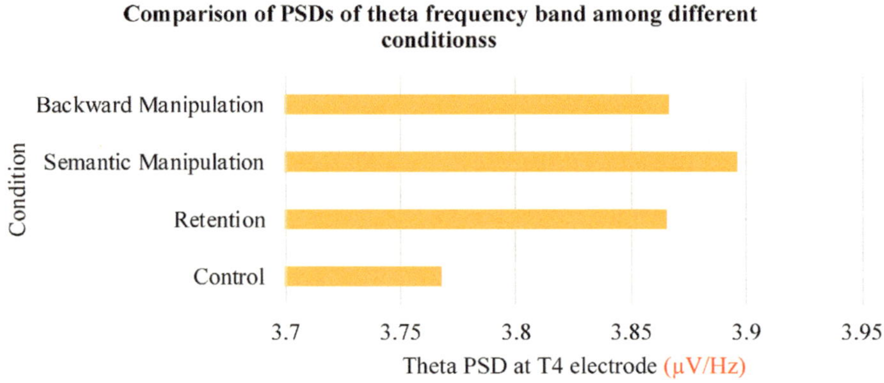

Figure 2.
Power spectral densities (PSDs) of theta frequency wave form in three memory conditions of retention, semantic manipulation and backward manipulation with raw EEG data being processed through BESS software where epochs (epoch length = 1000 ms) were separated for each trial [54 trials being part of delayed-match-to-sample (DMTS) task] and data was averaged separately respectively for each electrode for each condition (FP1, FP2, F7, F3, AFz, Fz, F4, F8, T3, C3, Cz, C4, T4, P7, P3, Pz, P4, P8, O1, O2 electrodes were selected). ERS as evinced through enhanced PSD (increase in mean amplitude power in sq. microvolts), was observed in theta wave-form in all three conditions/manoeuvres of retention (FZ, F3, F4, F7, T3, T4) semantic forward information processing (FP1, FP2, AFZ, FZ, F3, F4, F7, F8, C3, P7, T3, T4) and backward information processing (FP1, FP2, AFZ, FZ, F3, F4, F7, F8, C3, T3, T4) of EEG electrode pairs and on comparative evaluation with basal EEG time-series run along said EEG electrode pairs, significant difference in PSD could be appreciated only along T4 EEG electrode pair in conditions of retention ($p = 0.05$), semantic manipulation ($p = 0.05$) and backward manipulation ($p = 0.01$) by using one way ANOVA at 5% level of significance.

The assessment of power-spectral density of EEG signals from our laboratory paved the way for appreciation of closely intertwined intricate dance of *ERS/ERD along the coordinates of space and time that probably seems to be the flip-flop switch for the flow of corporeal and legible information* (**Figure 3**) [49]. The ERS of theta wave-form with significantly appreciable change in Power Spectral Density (PSD) at EEG electrode pair of T4 (**Figure 1**) along with concomitant ERD of alpha wave form skewing onto left hemisphere lateralisation of neurophysiological processes [as exemplified by Oblique Lateral Asymmetry Index (LAI)] during the select conditions of retention, semantic manipulation and backward manipulation (**Figure 2**) is yet another example of concomitant stimulatory and inhibitory dedicated neuronal pools that evolve during and are responsible for the stimulus-specific adequate response. More likely, the looped fractals of neuronal pools (modules) of on-centre ERS theta wave form, on-centre ERD alpha and offsurround ERS alpha or off-centre ERS alpha and onsurround ERD alpha have a tendency to self-iterate that tends to etch the stochastic trajectory along the Human Mind Phase-Space.

The characteristic of temporal distribution of *ERS/ERD PSD* along the run of EEG time-series was evaluated in our laboratory and during the **DMTS task, the temporal distribution across *two frequency bands of theta and alpha*** was accessed to assess neuronal oscillatory activities during **WM tasks** across select cortical regions [50–52] and to assess modular memory facets and processes that entrain dedicated self-iterating fractals of neuronal pools in human brain resulting in memory consolidation processes concluding into **language acquisition, manipulation and comprehension processes.**

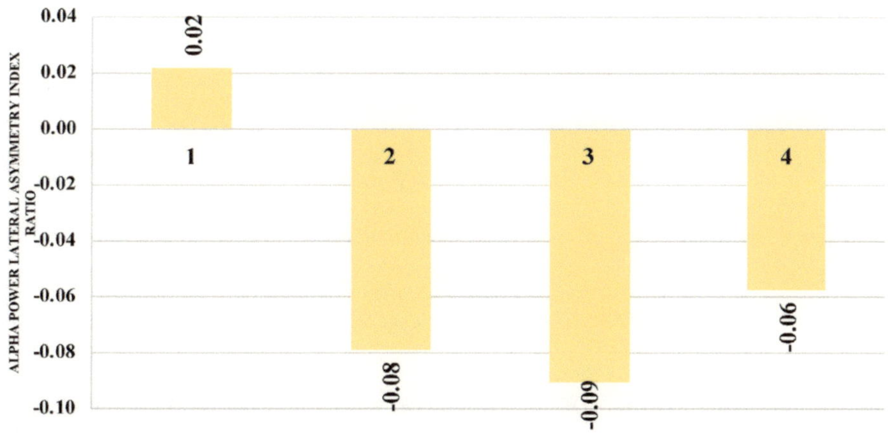

Figure 3.
Portrays Lateral Asymmetry Index (LAI) ratio of different conditions of Alpha Frequency Band at P3-P4 EEG electrode site. Significant difference could be appreciated at parietal region of P3-P4 EEG electrode pair along alpha frequency band using T-test with p = 0.004, p=0.002 (p < 0.05) at 5% level of significance with left hemisphere lateralisation (skewed neurophysiological processes) during retention condition and semantic manipulation condition, respectively. In backward manipulation condition, significant difference could be appreciated at additional parietal EEG lead pairs of P7-P8 besides P3-P4 with p = 0.02 in both electrode pairs (p < 0.05) at 5% level of significance with left hemisphere lateralisation. 1 = Control Condition, 2 = RetentionCondition, 3 = Semantic Condition, 4 = Backward Condition. LAI = [P (left) - P (right)]/[P (left) + p (right)].

These chunks of information or memory codes might generate a **particular patterned rhythm** which later during **retrieval** of information from dedicated neural networks might follow the phenomena of **pattern matching** during its response for same memory inputs.

Figures 4 and **5** depicts *ERD/ERS percentage change {ERD% = (Actual Power-Reference Power)/Reference Power × 100} of Power Spectral Densities of Alpha frequency wave form* when compared among male and females in *Retention Condition* where significant difference could be seen at CZ, P8, T4 electrode sites and **Figure 5** displays results of *ERD/ERS alpha activity* in *Semantic Condition*, exhibiting significant differences along CZ, P4, T4 EEG electrode sites. The common denominator appreciates the intricate interwoven ***Off-Centre α ERS (Central)-On-Surround α ERD Neural Dynamics*** as could be deduced and envisioned from observations of **Figures 4** and **5** that seem to be intertwined and interlocked through observations of findings of **Figures 2** and **3**, with the self-iterating trajectorial pathways of the

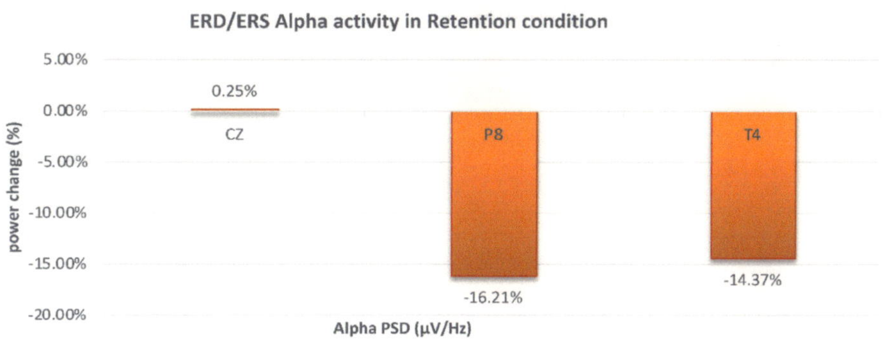

Figure 4.
Depicts ERD/ERS percentage change [ERD% = (Actual Power-Reference Power)/Reference Power × 100] of Power Spectral Densities of Alpha frequency wave form when compared among males and females in Retention Condition where significant difference could be observed at CZ, P8 and T4 EEG electrode sites.

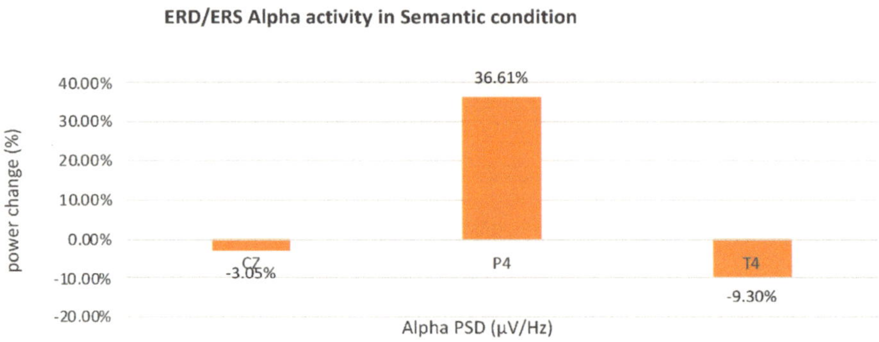

Figure 5.
Illustrates ERD/ERS percentage change {ERD% = (Actual Power-Reference Power)/Reference Power × 100} of Power Spectral Densities of Alpha frequency wave form when compared among male and females in Retention Condition where significant difference could be seen along CZ, P4, T4 EEG electrode sites.

looped alpha and theta wave forms through respective precepts of Event-Related Synchrony (ERS) and Event-Related Desynchrony (ERD).

In this context and with the characteristically patterned observations data from the present study [*Neuropsychological Trends* in print] the precepts of *Neural Dynamics of Working Memory Model* has been conceptualised as:

- **The Retention** Precept singularly involves Theta ERS along temporal regions with antecedent etched LAI (Lateral Asymmetry Index) Alpha ERD along fractal neuronal networks of parietal region.

- **The Semantic (Forward) Processing** appreciating the relevance of ascendance (increasing quantal framework) observes a similar patterned and looped Theta ERS (temporal region) with LAI Alpha ERD along parietal regional precept of fractal neuronal networks.

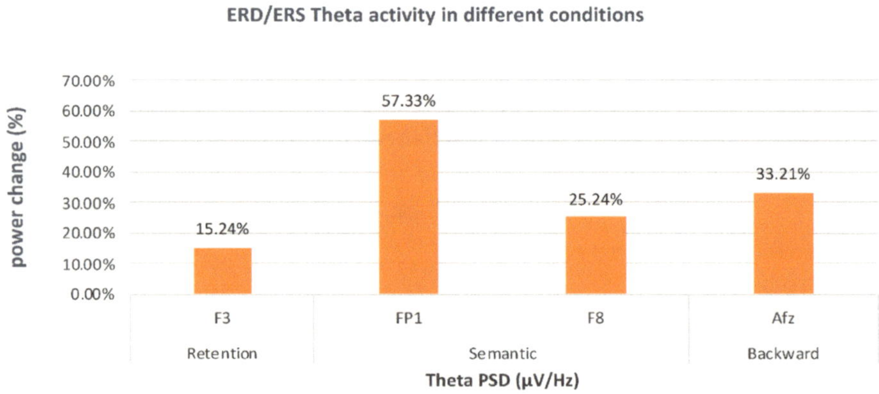

Figure 6.
ERD/ERS percentage change of Power Spectral Densities (PSDs)of Theta frequency wave form when compared among male and females in all the three conditions where significant difference could be appreciated at F3 in Retention condition; at FP1,F8 in Semantic manipulation condition and at AFz in Backward manipulation condition and results could reflect same as Alpha frequency wave form that females outperformed in the visuo-spatial DMTS task compared to males using T-test at 5% confidence level. The contextual inference from the present study in relation to the above stochastic phase-space trajectory of ***Off-Centre α ERS (Central)-On-Surround α ERD-On Surround θ ERS*** *document a significantly enhanced PSD values of said trajectorial path in females as compared to that observed in males, endowing the female gender with neurophysiologically efficient neural dynamics of working memory.*

- **The** ***Backward Processing*** *in Memory warrants a similar Theta ERS-LAI Alpha ERD looping along temporal and parietal region (with additional inputs from parietal areas).*

The hypothesis posited is that the concept of *Neural Dynamics of Working Memory Model* reflects as under:

- *The Theta Frequency Wave-Form ERS along temporal region* with concomitant and antecedent *LAI Alpha Frequency wave-form ERD along parietal terrain* characterise the ***Retention and Semantic Forward Information Processing*** *Precepts* and

- ***The Backward Information Processing Precept*** exemplified through *Theta Frequency Waveform ERS along temporal area* with concomitant and antecedent *LAI of Alpha Frequency Waveform ERD along extended parietal region* (**Figure 6**). The contextual inference from the present study (refer **Figures 1–6**) in relation to the above stochastic phase-space trajectory of ***Off-Center α ERS (Central)-On-Surround α ERD-On Surround θ ERS*** *document* a significantly enhanced PSD values of said trajectorial path in females as compared to that observed in males, endowing the female gender with neurophysiologically efficient *neural dynamics* of working memory.

It seems that there are two aspects of processing of LTM in terms of mean PSD and LAI along theta and alpha frequency waveforms.

- ***The Skewed Theta frequency waveform ERS along right temporal region*** **during** ***retention, semantic forward information processing and backward information processing conditions*** along with

- The Alpha frequency waveform ERD with Lateral Asymmetry Index spreading through parietal region during retention and semantic forward information processing conditionis suggestive of *laterality restricted looping of ERS-ERD console within the left hemisphere*.

The precept of ***Hemispheric Encoding/Retrieval Asymmetry (HERA)*** so documented had been first hypothesised by Tulving et al. [53] supported by Nyberg et al. [54] as well that advocates the premise of **preferential and skewed involvement of left hemisphere in semantic (algorithmic non-linear neural information flow) retrieval and encoding** whereas **right hemisphere seems to be more involved with the episodic retrieval.**

The visual sensory inputs/information so perceived in the form of varied protocols of ***Delayed Matched to Sample Task (DMTS)*** is essentially relayed to primary visual cortex underlying EEG occipital region electrode pairs where information is processed. Primary visual cortex (V17) [55] subserves the qualia of perception and visual association areas (V18, 19) [56]. Lisman and Jensen [57] concluded **the process of recognition through** ***patterned-matching of the gamma-burst, alpha-theta waveforms looping or the bootstrapping (piggy-back riding) of gamma burst onto alpha-theta combine waveforms.*** The visual inputs as a part of visuo-spatial *DMTS* are perceived by occipital region and it has been modelled [58] that such visual impulses are then translated and transmogrified into auditory impulses in the differently-abled angular gyrus (anterolateral region of parietal lobe, near the superior edge of temporal lobe and immediately posterior to the supramarginal gyrus), a feature that could be observed as *increase in the* *amplitude (ERS) of Theta Frequency wave-form in EEG*. The visual–auditory interface impulse is then transferred onto Wernicke's area/auditory neural codes (Brodmann area 22, superior temporal gyrus) in order to appreciate and decode the semantics of visuo-auditory interface impulse perceived as symbols, letters, words and matching sounds accordingly [59].

ERS of Theta Waveform so evolved by interacting stimulus-locked dedicated neuronal pools *with* ***ERD of Alpha waveform functionally and neurophysiologically representing the dedicated reverberating mirror neuronal pool*** *system seems to be representing* ***the working model of Human Memory-Language***. It seems that the generation of language shapes into the virtual stochastic phase-space of human mind through the help of **reverberating Lateral Asymmetry of** ***Alpha wave-form ERD, representative of Mirror Neurone System (MNS)***. Previous studies have reported that *Alpha ERD during motor response in a WM task has been interpreted as the preparation of a movement-specific motor task* but does not reflect processing for the specific task itself [43, 60]. The alpha ERD in the sensorimotor system may buttress the concept of **a preparatory role of alpha ERD**. Alpha ERD had been also posited even during anticipation of an event [61], again emphasising the role of preparation for a motor response. In this background, the *role of alpha ERD* could be perceived as developing a preparatory *schema intricately interwoven with the Mirror Neurone System (MNS) creating and evolving an image (an alter-image in the stochastic phase-space of Human Mind)* during the ensuing encoding interval.

The findings of ***ERS in theta wave-form*** *with a significant change in PSD along select EEG electrode pairs a recent study from our laboratory* have also been reported by [62, 63], though Burke et al. [21] and [46, 47] could not observe such a patterned and locked differential EEG theta wave-form PSD during the manoeuvres of retention, semantic and backward manipulation and hypothesised a possibility of contextual overlapping between encoding and retrieval tasks.

The above documentation of ***ERS Theta Frequency waveform bootstrapping*** *with concomitant* ***ERD of Alpha Frequency waveform*** *seem to evolve an envelope of Working Memory that translates into a comprehensible means of communication, Language*. The interplay between these frequency wave-form forms the ground of working memory which is thought to be an important constituent component instrumental in language acquisition, comprehension and manipulation. The amount of information/memory inputs restricted by day-to-day working memory might be useful and can be considered as the focus for processing and acquisition of language e.g., semantics of letters and words (positioning and placement), syntactics of words (reproducible neurodynamically grammatically cogent disposition/sequence), word frequency, plausibility, discourse context, intonational information, etc.

The processing of letters or words in the form of memory inputs give an insight into the underlying neuro-physiological processing and neural dynamics responsible for the evolution and progression of the evolved phenomena of written and spoken language that make use of semantic and episodic memory. The *EEG Power Spectral Densities* ***(PSDs) of alpha frequency band during semantic memory and information processing*** and *the* ***PSDs of theta frequency band during episodic memory and information processing*** that follow separate paths in their nativity could be responsible for holding relevant information across coordinates of space and time (freezing the flow of space and time in the process) providing a gateway for synthesis of a structured and evolved system of communication, known as *language*.

The above observations create the platform for an integrating function and role of principles of *Working Memory in generation and evolution of a synthesised and coordinated communication system as outlined by the structured Language of Human Mind.*

8. The arena of language acquisition: probable neural substrates and signature

The major debate regarding *neural substrates underlying language acquisition (inclusive of the capacity to detect phonetic distinction and develop language – specific phonetic capacity and acquire legible, valid and comprehensible words)* lies in the belief if nativist (innate rather than acquired) domain – specific dedicated neural mechanism(s) operate exclusively on linguistic data, wherein the neural architecture is decided beforehand for an individual in acquisition of language or general learning mechanism(s) contribute to such an evolved mechanism of spoken and written language. The nativist approach posits the universal capacity to detect differences in phonetic contrasts in all languages. It has further been hypothesised from ERP studies that the response profile of *Human Mind* in terms of ERPs that are locked in space and time to varied phonetics is a significantly important component contributing to elementary building blocks of language and initial language phonetic learning is an essential pathway to learning.

Hence, it seems that **the fine dance of *ERS Theta Frequency wave-form* observed at temporal EEG lead pair *closely looped with LAI of Alpha Frequency wave-form ERD* seem to evolve a synthesising envelope of *Working Memory*** that translates into comprehensible means of communication, ***Language***. The *theta and alpha frequency waveforms* with the available resources, the interplay between these frequency waveforms, initiate the ground of working memory which seemingly is hitched-hiked onto **language acquisition, comprehension and manipulation**. The dynamical power spectral interplay of *theta and alpha frequency waveforms* along the coordinates of space and time during the *Working Memory* tasks of *retention, semantic (forward processing) and backward processing* seem to form the *gateway of primacy* opening the portal of algorithmic flow of neural information so needed for the neurocognitive primacy of language. The amount of information/memory inputs constraint by quotidian working memory might be utilitarian and can be considered as cynosure for processing and acquisition of language e.g., semantics of letters and words, syntactics of words, word frequency, plausibility, discourse context, intonational information, to name some of the intricate and fascinating nuances.

Subsequently, it is conceived that self-iterating fractal of interacting *ERD* and *ERS* through respective frequency waveforms *theta* (θ) and *alpha* (**α**) *waveforms* is construed with θ*waveform band singularity of ERS* across frontal and midline regions with antecedent **α***ERD* across respective mirror neurone system domain along with**α***ERS* at central region. The singularity of *ERS* denotes a preferential and categorical *inhibition gateway* and an *ERD* represents an event related and locked gateway to stimulatory/excitatory neuronal architectonics presumably responsible for *stimulus-locked and adequate neural response*. The fine and intricate interplay of θ *ERS (frontal and midline areas fine-tuned excitation)*, **α** *ERD (parietal and temporal floral activation) and* **α** *ERS (central selective inhibition)* evolves the self-evolving florid landscape of an **ERD on-centre and ERS off-surround loci along with an ERS off-centre and ERD on-surround**. The evolution of *frontal and midline excitatory* θ *ERS* along stochastic phase-space trajectory is a reflection of an evolving *fractal self-iterating excitatory gateway with antecedent fine-tuned channelisation of attentional mechanisms onto the stimulus/event restricting extraneous interfering neural mechanisms in the process*. The florid **α** *ERD* is representative of an evolving *excitatory stochastic phase-space trajectory dynamically mirroring the functional Mirror Neurone System (MNS)* responsible for algorithmic information flow onto subsequent *MNS* along with antecedent *central selective inhibition through* **α** *ERS inhibiting interfering contrivances (an example of* **α** *ERS Off-Centre with* **α** *ERD on-surround with*

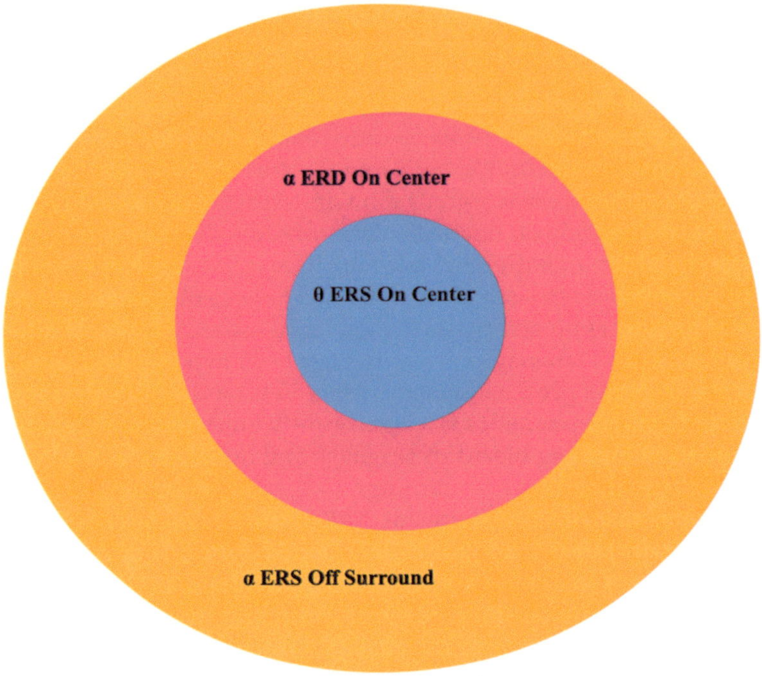

Figure 7.
The theta (θ) ERD on centre, alpha (α) ERD on centre and alpha (α) ERS off surround model.

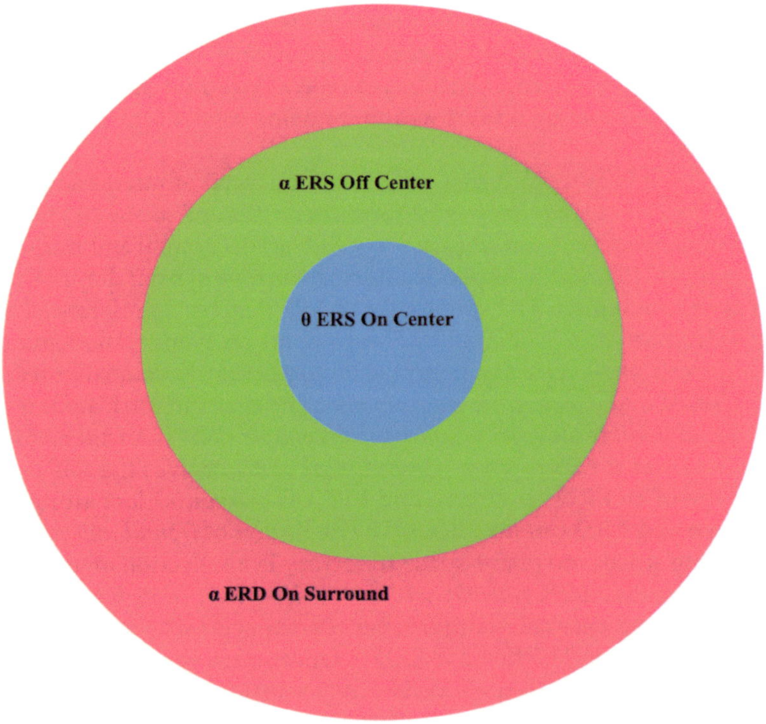

Figure 8.
The theta (θ) ERS on centre, alpha (α) ERS off centre and alpha (α) ERD on surround model.

θ *ERS on-surround).* These self-iterating fractal architectonics of *central inhibitory, Off-Centre* **α** *ERS, surround excitatory on-surround* **α** *ERD and on surround* θ *ERS*, representative of *interwoven PSD singular* ***Off-Centre*** **α** ***ERS (Central)-On-Surround*** **α** ***ERD-On Surround*** θ ***ERS*** *phenomenology* seem to define the qualia and quanta of underlying neural mechanisms of *working memory* (**Figure** 7).

The contextual inference in relation to *stochastic phase-space trajectory of* ***Off-Centre*** **α** ***ERS (Central)-On-Surround*** **α** ***ERD-On Surround*** θ ***ERS*** document a neurophysiologically efficient *neural dynamics of working memory* (**Figure 8**) [49].

The above model envisages a self-iterating fractal of *θ ERS On-Centre along with α ERD On-Centre and α ERS Off-Surround* mirrored along *θ ERS On-Centre, α ERS Off-Centre and α ERSD On-Surround, the so-called EEG micro states that tend to oscillate through the execution of the respective cognitive manoeuvre and these self-iterating fractals of lateral asymmetry index (LAI) of alpha (α) ERD and ERS along with theta (θ) ERS tend to open the gateway/portal of effective cognitive network.*

In this connectome, the *Human Mind* is envisaged as an esoteric concept that probably represents a logical synthesis of functional *mass* and *energy*, so represented by the characteristically patterned modulation and flow of the neurally coded information.

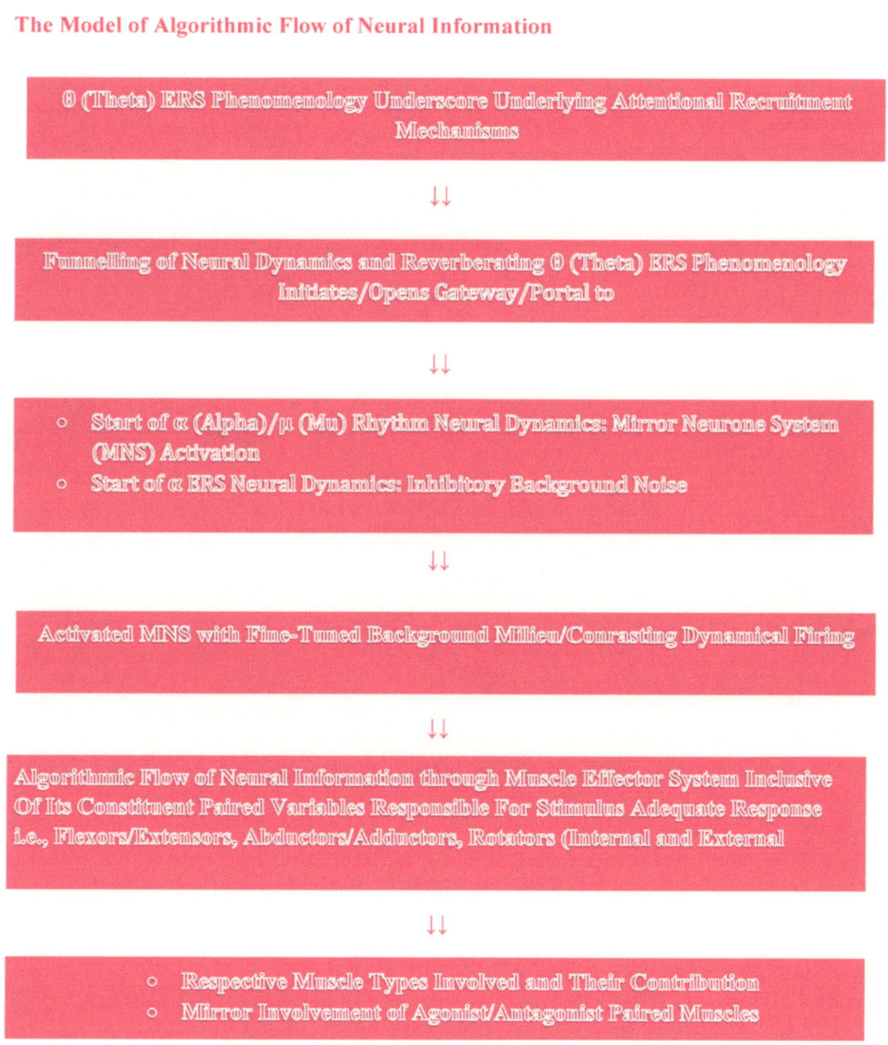

In conclusion, the neural architectonics subserving *language* seem to evolve across the self-iterating fractal features of phenomenology of *on-centre/off surround and off centre/off surround of ERD and ERS represented through electroencephalographic frequency waveforms of* θ*ERS,* α*ERD and* α*ERS* that synthesise and evolve the fine-tuned cognate neural mechanisms that evolve into the structured means of communication, *language.*

Contributors

Amitabh Dube, M.D., Umesh Kumar, M.D., Bhoopendra Patel, M.D., Lubaina Jetaji, M.Sc., Kapil Gupta, M.D., Jitendra Gupta, M.D., Sanjay Kumar Singhal, M.D., Kavita Yadav, M.Sc., Ph.D., Shubha Dube, Ph.D

Acknowledgements

The authors are greatly indebted to **Late Professor (Dr.) Ashok Panagariya**, a *Neuroscientist of International acclaim*, whose piercing insight and vision had been and will remain *The Fountainhead* of the ongoing work on *Modelling of Human Mind*.

Author details

Amitabh Dube[1*], Umesh Kumar[2], Kapil Gupta[1], Jitendra Gupta[1], Bhoopendra Patel[3], Sanjay Kumar Singhal[1], Kavita Yadav[1], Lubaina Jetaji[1] and Shubha Dube[4]

1 Physiology, S.M.S. Medical College and Attached Hospitals, Jaipur, Rajasthan, India

2 Physiology, Government Medical College, Kota, Rajasthan, India

3 Physiology, AIIMS, Bilaspur, Himachal Pradesh, India

4 Human Development, University of Rajasthan, Jaipur, Rajasthan, India

*Address all correspondence to: amitabhdube786@gmail.com

References

[1] Fingelkurts A and Fingelkurts A. 2004. Making complexity simpler: multi variability and metastability in the brain. International Journal of Neuroscience, 114 (7), 843-862.

[2] Abraham, F.D. (1995). Dynamics, bifurcations, self- organization, chaos, and mind. In R. Robertson and A. Combs (Eds.). Proceedings of The Society for Chaos Theory and the Life Sciences. Lawrence Erlbaum

[3] Freeman, W.J. (1991). The physiology of perception. Scientific America, *264, 78-85.*

[4] Dube A, Kumar A, Gupta K, Vyas P, Boolchandani D and Sonania R. 2009. Chaotic Neural Dynamics as evinced from scalp electroencephalography (EEG). 13th Congress of the European Federation of Neurological Societies (EFNS), September 12 to 15, 2009. European Journal of Neurology, Volume 16, Supplement 3, September 2009, pg 451

[5] Kampis, G. (1991). Self-modifying systems in biology and cognitive science. New York: Pergamon.

[6] Krause, C.M. (2003). Brain electric oscillations and cognitive processes. In: Hugdahl, K. (Ed.), Experimental Methods in Neuropsychology. Neuropsychology and Cognition. Kluwer Academic Publishers, Boston, pp. 111–130.

[7] Buzsáki, G.; Draguhn, A. (2004). Neuronal oscillations in cortical networks. Science, 304, 1926–1929.

[8] Jasper, H.H. (1958) The Ten-Twenty Electrode System of the International Federation. Electroencephalography and Clinical Neurophysiology, 10, 371-375.

[9] Skarda, C.A. and Freeman, W.J. (1987). How brains make chaos in order to make sense of the world. Behavioral and Brain Sciences, *10 (2), 161-195.*

[10] Basar, E., Basar-Eroglu, C., Karakas, S., Schurmann, M. (1999). Are cognitive processes manifested in event-related gamma, alpha, theta and delta oscillations in the EEG? Neurosci. Lett. 259, 165–168.

[11] Ramachandran VS and Oberman LM. 2007. "*Broken Mirrors: A special theory of Autism*". Scientific American, 17 (2s), 63-69.

[12] Dube A, Yadav K, Gupta A and Yadav Y. 2012. Neurophysiological perspectives of electroencephalography in children with attention deficit hyperactivity disorder (ADHD). Neurophysiologie Clinique Clinical Neurophysiology, 42 (1–2), 68.

[13] Krause, C.M., Åström, T., Karrasch, M., Laine, M., Sillanmäki, L. (1999). Cortical activation related to auditory semantic matching of concrete versus abstract words. Clin. Neurophysiol., 110, 1371–1377.

[14] Krause, C.M., Sillanmäki, L., Koivisto, M., Saarela, C., Häggqvist, A., Laine, M., Hämäläinen, H., (2000). The effects of memory load on event-related EEG desynchronization and synchronization. Clin. Neurophysiol. 111, 2071–2078.

[15] Bloom, J. L., & Anneveld, M. (1982). An electrode cap tested. Electroencephalography and Clinical Neurophysiology, 54, 591–594.

[16] Egner, T., & Gruzelier, J. H. (2001). Learned self-regulation of EEG frequency components affects attention and event-related brain potentials in humans. NeuroReport: For Rapid Communication of Neuroscience Research, 12(18), 4155–4159.

[17] Luck, S. J. (2014). An Introduction to the Event-Related Potential Technique, Second Edition. Cambridge, MA: MIT Press.

[18] Beres. (2017).Time is of the Essence: A Review of Electroencephalography (EEG) and Event-Related Brain Potentials (ERPs) in Language Research. Applied Pyschophysiology and Biofeedback Volume 42, Issue 4, pp. 247–255.

[19] Berger, B.; Omer, S.; Minarik, T.; Sterr, A.; Sauseng, P. (2015) Interacting Memory Systems—Does EEG Alpha Activity Respond to Semantic Long-Term Memory Access in a Working Memory Task? Biology, *4*, 1-16.

[20] Burke JF, Long NM, Zaghloul KA, Sharan AD, Sperling MR, Kahana MJ. (2014) Human intracranial high-frequency activity maps episodic memory formation in space and time. NeuroImage., 2, 834–843.

[21] Burke JF, Zaghloul KA, Jacobs J, Williams RB, Sperling MR, Sharan AD, Kahana MJ. (2013) Synchronous and asynchronous theta and gamma activity during episodic memory formation. Journal of Neuroscience, 1, 292–304.

[22] Doppelmayr, M., Klimesch, W., Stadler, W., Pöllhuber, D., & Heine, C. (2002). EEG alpha power and intelligence. Intelligence, 30, 289–302.

[23] Kuhl PK. (2010). Brain mechanisms in early language acquisition. Neuron, 67 (5), 713-727.

[24] Postle, B.R. (2006). Working memory as an emergent property of the mind and brain. Neuroscience, 139, 23–38.

[25] Baddeley, A. (2003). Working memory: looking back and looking forward. Nat. Rev., Neurosci. 4, 829–839

[26] Baddeley, A. (2007). Oxford psychology series: Vol. 45. Working memory, thought, and action. Oxford University Press.

[27] Jensen, O., Tesche, C.D. (2002). Frontal theta activity in humans increases with memory load in a working memory task. Eur. J. Neurosci. 15, 1395–1399.

[28] Kahana MJ, Sekuler R, Caplan JB, Kirschen M, Madsen JR. (1999). Human theta oscillations exhibit task dependence during virtual maze navigation. Nature 399:781–784.

[29] Klimesch et al. (1994). Episodic and semantic memory: an analysis in the EEG theta and alpha band. Electroencephalography and Clinical Neurophysiology, 91(6), 428-441.

[30] Klimesch et al. (1998). EEG alpha and theta oscillations reflect cognitive and memory performance: a review and analysis. Brain Res Brain Res Rev, 29 (2–3), 169-195.

[31] Griesmayr B., Gruber W., Klimesch W. (2010). Sauseng P. Human frontal midline theta and its synchronization to gamma during a verbal delayed match to sample task. Neurobiol. Learn. Mem., 93, 208–215. doi: 10.1016/j.nlm.2009.09.013

[32] Lisman, J. E., & Idiart, M. A. P. (1995). Storage of 7±2 short-term memories in oscillatory subcycles. Science, *267*(5203), 1512–1515.

[33] Klimesch et al. (1997). Brain oscillations and human memory: EEG correlates in the upper alpha and theta band. Neuroscience Letter, Nov 28;238 (1-2):9-12.

[34] Klimesch et al. (2007). EEG alpha oscillations: the inhibition-timing hypothesis. Brain Res Rev. Jan;53(1): 63-88. Epub 2006 Aug 1.

[35] Klimesch et al. (2011). Alpha oscillations and early stages of visual

encoding. Front Psychology, May 31;2: 118.

[36] Klimesch et al. (1996). Memory processes, brain oscillations and EEG synchronization. Int J Psychophysiol. 24 (1–2):61-100.

[37] Röhm D, Klimesch W, Haider H, Doppelmayr M. (2001). The role of theta and alpha oscillations for language comprehension in the human electroencephalogram. Neurosci Lett, 310(2–3), 137-140.

[38] Pfurtscheller, G., Klimesch, W. (1990) Topographical display and interpretation of event-related desynchronization during a visual-verbal task. Brain Topogr **3,** 85–93.

[39] Pfurtscheller, G., Aranibar, A. (1977). Event-related cortical desynchronization detected by power measurements of scalp EEG. Electroencephalogr. Clin. Neurophysiol. 42, 817–826.

[40] Pfurtscheller, G., Lopes da Silva, F. H. (1999). Event-related EEG/MEG synchronization and desynchronization: basic principles. Clin. Neurophysiol. 110, 1842–1857.

[41] Sauseng, Paul & Klimesch, Wolfgang & Schabus, Manuel & Doppelmayr, Michael. (2005). Fronto-parietal EEG coherence in theta and upper alpha reflect central executive functions of working memory. International journal of psychophysiology: official journal of the International Organization of Psychophysiology, 57, 97-103.

[42] Sauseng P., Klimesch W. (2008). What does phase oscillatory brain activity tell us about cognitive processes? Neurosci. Biobehav. Rev. 32, 1001–1013.

[43] Wianda, Elvis & Ross, Bernhard. (2019). The roles of alpha oscillation in working memory retention. Brain and Behavior, 9, 10.1002/brb3.1263.

[44] Stipacek, A., Grabner, R.H., Neuper, C., Fink, A., Neubauer, A.C. (2003). Sensitivity of human EEG alpha band desynchronization to different working memory components and increasing levels of memory load. Neurosci. Lett., 353, 193–196.

[45] Sarnthein J., Petsche H., Rappelsberger P., Shaw G. L., von Stein A. (1998). Synchronisation between prefrontal and posterior association cortex during human working memory. Proc. Natl. Acad. Sci. U.S.A. 95, 7092–7096. 10.1073/pnas.95.12.7092.

[46] Hanslmayr S, Staudigl T. (2013) How brain oscillations form memories-a processing based perspective on oscillatory subsequent memory effects. NeuroImage, 5, 326–334.

[47] Hanslmayr S., Pastötter B., Bäuml K. H., Gruber S., Wimber M., Klimesch W. (2008). The electrophysiological dynamics of interference during the stroop task. J. Cogn. Neurosci., 20, 215–225.

[48] Nigbur R., Cohen M. X., Ridderinkhof K. R., Stürmer B. (2012). Theta dynamics reveal domain-specific control over stimulus and response conflict. J. Cogn. Neurosci, 24 (5), 1264-1274.

[49] Lubaina, et al. (2021). Neuropsychological Trends. In print.

[50] Guillery, R.W. (2000), Brodmann's'Localisation in the Cerebral Cortex'. Translated and edited by Laurence J. Garey. (Isbn 1 86094 176 1.) London: Imperial College Press. 1999.. Journal of Anatomy, 196, 493-496.

[51] Thut G, Nietzel A, Brandt SA, Pascual-Leone A. (2006) α-Band electroencephalographic activity over

occipital cortex indexes visuospatial attention bias and predicts visual target detection. J Neurosci, 26, 9494–9502. doi:10.1523/JNEUROSCI.0875-06.2006 pmid:16971533.

[52] Van Ede F, de Lange FP, Jensen O, and Maris E (2011). Orienting attention to an upcoming tactile event involves a spatially and temporally specific modulation of sensorimotor alpha- and beta-band oscillations. Journal of Neuroscience 31, 2016-2024.

[53] Tulving, E., Kapur, S., Craik, F.I.M., Moscovitch, M. and Houle, S. (1994) Hemispheric encoding/retrieval asymmetry in episodic memory: positron emission tomography findings, Proc. Nat. Acad. Sci. USA, 91, 2008–2011.

[54] Nyberg, L., Cabeza, R. and Tulving, E. (1995) PET studies of encoding and retrieval: the HERA model, Psychol. Bull. Rev., 3, 135–148.

[55] Broadmann K. (1909) Broadmann's: Localisation in the cerebral cortex. Third edition. Springer

[56] Sherman SM and Guillery RW. (2000). Exploring the thalamus. Brain, 124 (10), 2120-2121

[57] Lisman JE and Jensen O. (2013). The θ-γ neural code. Neurone, 77 (6), 1002-1016.

[58] Desmond, J. E., Gabrieli, J. D. E., Wagner, A. D., Ginier, B. L., & Glover, G. H. (1997). Lobular patterns of cerebellar activation of verbal working-memory and finger-tapping tasks as revealed by functional MRI. The Journal of Neuroscience, *17*(24), 9675–9685.

[59] Bogen JE, Bogen GM. (1976). Wernicke's region–Where is it? Ann N Y Acad Sci., 280, 834-843. doi: 10.1111/j.1749-6632.1976.tb25546.x. PMID: 1070943.

[60] Deiber, Marie-Pierre & Sallard, Etienne & Ludwig, Catherine & Ghezzi, Catherine & Barral, Jérôme & Ibañez, Vicente. (2012). EEG alpha activity reflects motor preparation rather than the mode of action selection. Frontiers in integrative neuroscience. 6. 59. 10.3389/fnint.2012.00059.

[61] Bastiaansen MC, Böcker KB, Cluitmans PJ, Brunia CH. (1999) Event-related desynchronization related to the anticipation of a stimulus providing knowledge of results. Clin Neurophysiol., 110(2), 250-260. doi: 10.1016/s0013-4694(98)00122-9. PMID: 10210614.

[62] Kopp F., Schröger E., Lipka S. (2006). Synchronized brain activity during rehearsal and short-term memory disruption by irrelevant speech is affected by recall mode. Int. J. Psychophysiol. 61, 188–203 10.1016/j.ijpsycho.2005.10.001.

[63] Payne L., Kounios J. (2009). Coherent oscillatory networks supporting short-term memory retention. Brain Res. 1247, 126–132.

Chapter 2

A Brief Summary of EEG Artifact Handling

İbrahim Kaya

Abstract

There are various obstacles in the way of use of EEG. Among these, the major obstacles are the artifacts. While some artifacts are avoidable, due to the nature of the EEG techniques there are inevitable artifacts as well. Artifacts can be categorized as internal/physiological or external/non-physiological. The most common internal artifacts are ocular or muscular origins. Internal artifacts are difficult to detect and remove, because they contain signal information as well. For both resting state EEG and ERP studies, artifact handling needs to be carefully carried out in order to retain the maximal signal. Therefore, an effective management of these inevitable artifacts is critical for the EEG based researches. Many researchers from various fields studied this challenging phenomenon and came up with some solutions. However, the developed methods are not well known by the real practitioners of EEG as a tool because of their limited knowledge about these engineering approaches. They still use the traditional visual inspection of the EEG. This work aims to inform the researchers working in the field of EEG about the artifacts and artifact management options available in order to increase the awareness of the available tools such as EEG preprocessing pipelines.

Keywords: Artifact, Artifact removal methods, EEG, EEG preprocessing, Muscular artifacts, Ocular artifacts, Preprocessing pipelines

1. Introduction

A signal is a function that conveys information about the behavior or attributes of some phenomenon [1]. On the other hand, information can be anything. A waveform can have multiple overlapping information in the same space–time. The signal in a waveform is subjective, it can be color for one and shape for the other. In electrophysiology, waveform under inspection can be separated into two as the signal of interest and noise. The signal can be electrocardiography (ECG), Electroencephalogram (EEG), or any other physiological signal, noise is any unwanted wave source interfering with the signal. If we consider EEG as the signal, it is recorded from the scalp by electrodes and consists of the overall electrical activities of neural populations and a contribution of glial cells [2]. EEG has a wide range of use in both clinical practice and engineering applications in medicine, particularly neurology, sleep, and epilepsy research.

2. Background

The EEG recording environment and subject related electrical activities during recording deteriorate the signal quality. Artifacts are undesired signals that may introduce changes in the measurements and affect the signal of interest [3]. EEG can be contaminated in frequency or time domain by artifacts that are resulted from internal sources of physiologic activities and movement of the subject and/or external sources of environmental interferences, equipment, movement of electrodes and cables [4]. Artifact types and sources are listed in the **Table 1**. External artifacts can be prevented by proper shielding, grounding cables, isolating and moving cables away from recording sites since they act as antennas during operation. On the other hand, internal or physiological artifacts are challenging for researchers because of their inclusion of signal or resemblance to the signals. The most important artifacts in a typical EEG recording are ocular electro-oculogram (EOG) artifacts and muscular (EMG) artifacts.

2.1 Ocular artifacts

Electrical potentials due to eye opening/closure, blinks, eyelid flutter and eye movements propagate over the scalp and produce hostile EOG artifacts in the recorded EEG. Eye movements are major sources of contamination of EEG. The origin of this contamination is disputable. Cornea-retinal dipole movement, retinal dipole movement and eyelid movement are the three main proposed causes of the eye movement related voltage potential [6]. The direction of eye movements affects the shape of the EOG waveform while a square-like EOG wave is produced by vertical eye movements and blinks which leads to a spike-shaped waveform [7]. Blinks

Artifact	Type	Source
Eye blink	Ocular	Internal/Physiological
Eye movement	Ocular	Internal/Physiological
REM Sleep	Ocular	Internal/Physiological
Scalp contractions	Muscle	Internal/Physiological
Glossokinetic artifact	Muscle	Internal/Physiological
Chewing	Muscle	Internal/Physiological
Talking	Muscle	Internal/Physiological
EKG	Cardiac	Internal/Physiological
Swallowing	Muscle	Internal/Physiological
Respiration	Respiratory	Internal/Physiological
Galvanic Skin Response	Skin	Internal/Physiological
Sweating	Skin	Internal/Physiological
Electrode movement	Instrumental	External/Extra-physiological
Electrode Impedence Imbalance	Instrumental	External/Extra-physiological
Cable movement	Instrumental	External/Extra-physiological
Electromagnetic coupling	Electromagnetic	External/Extra-physiological
Powerline	Electrical	External/Extra-physiological
Head movement	Movement	External/Extra-physiological
Body movement	Movement	External/Extra-physiological
Limbs movement	Movement	External/Extra-physiological

Table 1.
EEG artifact types and sources. Adapted from [4, 5].

which are attributable to the eyelid moving over the cornea, occurring at intervals of 1-10s, generate a characteristic brief potential of between 0.2 s and 0.4 s duration due to eyelid movement over cornea [8, 9]. The blinking artifact generally has an amplitude much larger than that of the background EEG [6]. It is advantageous to have a reference EOG channel during EEG recording for the cancellation of ocular artifact from EEG activity [3].

2.2 Muscular artifacts

Electrical activity on the body surface due to the contracting muscles are recorded via Electromyogram (EMG) [3]. Since independent myogenic activities of head, face and neck muscles are conducted through the entire scalp, it can be monitored in the EEG [10, 11]. The amplitude of this type of artifact is dependent on the type of muscle and the degree of tension [3, 12]. The frequency range of EMG activity is wide, being maximal at frequencies higher than 30 Hz [13, 14].

2.3 Cardiac artifacts

The electrical potential due to cardiac activity can exhibit itself in the EEG as ECG artifacts. Typical high frequency waveforms similar to EKG P-QRS-T shape are characteristics of EKG artifacts in EEG [15].

2.4 Other artifacts

Head, body and limb movements cause irregular high voltage artifacts. Artifacts can be produced by tremors in patients such as Parkinson disease and movement disorders. Changing patient position into a calm comfortable stable position helps reducing artifacts. Another prevention for respiratory related movement artifacts is to use a towel or a firm material support for the neck. The changes in the impedance or electrical potential between scalp and electrode may cause electrode artifacts. These can result from poor electrode contact, broken lead, electrolyte gel insufficiency. This type of artifact usually exhibits itself in sudden electrode pops. These electrode artifacts can be eliminated by using proper electrolyte gel, checking electrode impedance, changing the broken electrodes, and shifting the electrode position slightly.

3. Artifact handling methods

A typical EEG recording system is shown in **Figure 1**. At the heart of a recording setup is the biopotential amplifier. It should have high common mode rejection ratios, however it should not have high gains, this can saturate the signal due to large half-cell potentials at the electrodes. Unequal electrode impedances are major sources of common mode artifacts such as powerline.

Environmental artifacts can be eliminated by bringing the electrodes leads closer together, moving the electrodes and subject away from the noise sources, using single isolated earth for the whole setup, and shielding the cables, machines and artifact sources with a metal tape connected to the common earth. Moreover, the environmental conditions should satisfy the following requirements for proper recordings. These can be listed as, quiet atmosphere, comfortable temperature and humidity, controlled proper lighting, using a comfortable bed or chair, and separating the powerline of the EEG system from the other machines in the lab.

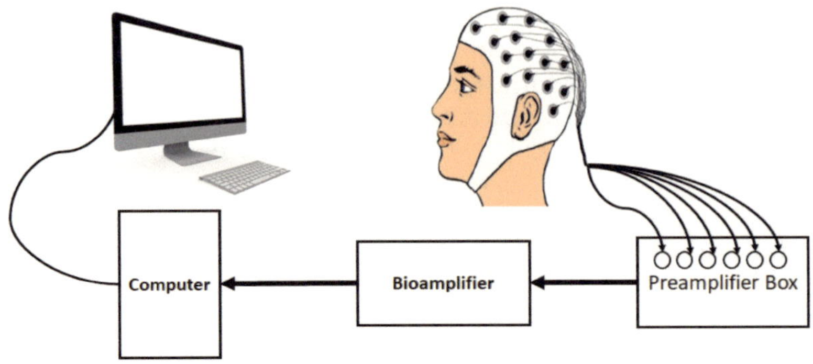

Figure 1.
EEG recording system and experiment setup.

3.1 Averaging methods to suppress ERP artifacts

Event Related Potentials (ERP) are electrical signals generated in response to internal or external events and they are recorded by EEG [16]. In evoked potentials, each stimulus produces an evoked potential embedded in EEG. However, since the ERP or evoked potential signals are generally subtle in EEG, averaging of many epochs are needed to make them distinguishable. An ensemble averaging method to enhance the ERPs was defined by [17]. This relies on the assumption that by synchronous averaging of each epoch, signal ERP amplitude adds constructively and EEG background noise diminishes destructively.

In ERP and evoked potential research, artifacts contaminate the final ensemble average signal of interest. One method to overcome this adverse effect is to benefit from a weighted averaging [18]. In weighted averaging technique each epoch is weighted inversely with the non-stationary noise maximum amplitude in the epoch. In [19], each trial's contribution to ensemble average is multiplied by a weight according to its correlation with the rest of the data. This factor is inversely related to its probability of being an artifact. For example, a large amplitude EEG is likely to be an artifact and the contribution factor for the trial involving large amplitudes will be low whereas the factor for a small amplitude EEG is high (**Figure 2**). Davila and Mobin [20] showed that weighted averaging of auditory EP has higher SNR than

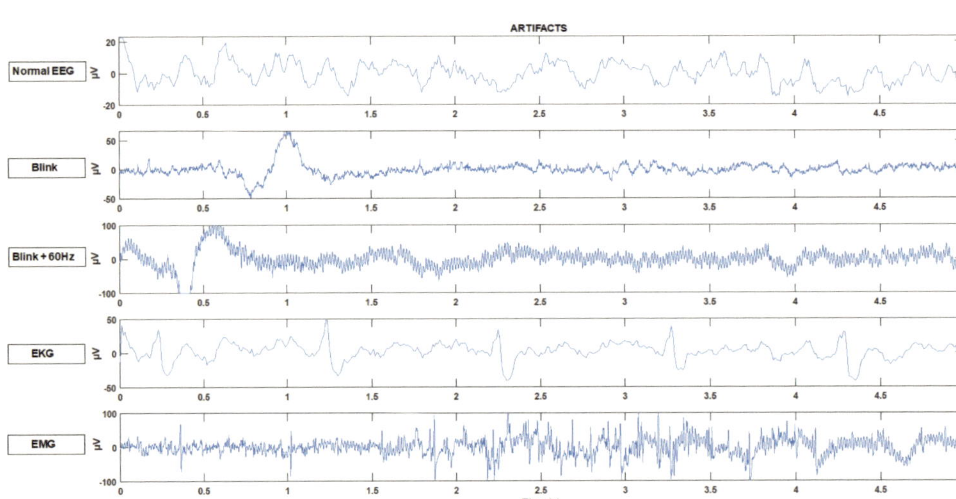

Figure 2.
Various EEG artifacts are shown.

conventional ensemble averaging. John et al. [21] studied the effects of such techniques as sample-weighted averaging, noise-weighted averaging, amplitude based artifact rejection, percentage based artifact rejection, and normal averaging on the steady state auditory evoked potentials. It concluded in favor of weighted averaging for better SNR of steady state responses. On the other hand, according to [22], weighted averaging underestimates the ERP signal amplitude. Determination of the optimal weighting factor is not straightforward and this limits the performance of the weighting averaging method. Mühler and Specht [23] developed a method called 'sorted averaging'. In sorted averaging, epochs are sorted with RMS values from small to large, since noisy artifactual epochs have large RMS values compared to low noise signals. The signal averaging is performed by addition of epochs from the low noise RMS to large RMS sorted order until a maximum peak of SNR^2 is obtained [24]. This eliminates the high RMS noisy epochs and yields a better ERP waveform. Compared to weighted averaging, sorted averaging had significantly higher SNR^2 [23].

Median averaging is another approach to ERP artifact handling and it is based on taking the median points of all the epochs and adding them to form a median average instead of classic mean average [25]. Some advantages of the median averaging are that; it elicits hidden signals more clearly and it is not affected by infrequent large artifacts that much compared to mean averaging [25]. Özdamar and Kalayci [26] supported the advantages of median averaging over the conventional mean averaging in a study on the ABR signals. Median averaging is an efficient way to remove adverse effects of the outliers on the final averaged signal, yet it also removes the valuable data in the outliers causing significant loss of information [27, 28].

3.2 Artifact handling methods for EEG

Artifact avoidance, artifact rejection, manual rejection, automatic rejection, and artifact removal are the common methods to deal with artifacts [29]. Although it seems a simple solution to cancel EOG and EMG artifacts by instructing subject to avoid blinking or movement, it can result in change of amplitudes in evoked potentials as well as the additional cognitive load [29–31]. On the other hand, artifact rejection or manual rejection may require a person dedicated to this purpose of eliminating artifacts visually one by one in an EEG. Moreover, the artifact detection by an expert may be subjective, tedious, and time consuming. In addition, it can not be applicable to online removal [3]. However, automatic rejection can automate this artifact rejection procedure but it can eliminate non-artifact signals if not properly tuned. The automatic rejection of artifact containing EEG can depend on artifact amplitude based or EEG segment RMS based artifact detection and rejection. An example of a simple blink artifact removal is depicted in **Figure 3**. Since blinks have low frequency content compared to EEG, by low pass filtering, EEG can be reduced while blink artifact still remains at a high voltage level. Thus, an amplitude threshold based artifact rejection can be applied. As seen from **Figure 3**, red traces are the EEG and blue are the low pass filtered EEG signal. While a simple artifact rejection (without low pass filtering) using a threshold of 20 μV will produce false positives (red traces over 20 μV), in the low pass filtered EEG these false positives are prevented.

Usually one or two channels are dedicated to detect EOG artifacts. There are two widely used procedures for EOG artifacts, first EOG rejection where EEG trials with EOG artifacts having VEOG greater than a preset threshold are omitted, and second EOG correction where the effect of eye movement is tried to be removed from EEG [6].

Artifacts can distort the EEG in a way that the electrophysiologists or physicians can be misled in their clinical interpretation [32]. This makes artifact removal critical in the pre-processing phase prior to analysis. There are many methods to

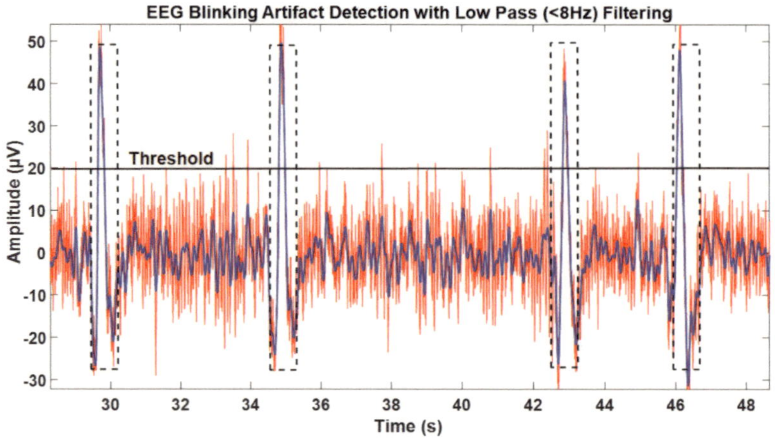

Figure 3.
Low pass filtering based EEG blink rejection. Red is raw EEG, blue is low pass filered EEG with 6th order Butteworth low pass filter at 8 Hz cut off. The detected artifact containing EEG epochs are shown in dashed rectangles.

remove artifacts such as Artifactual Segment Rejection, Filtering, Wiener filtering, Adaptive Filtering, Time-Frequency Representation, Wavelet Transform, Discrete Wavelet Transform (DWT), Adaptive Noise Cancelation (ANC), Wavelet Packet Transform (WPT), Kalman Filtering, Linear Regression, Blind Source Separation (Principal Component Analysis (PCA), Independent Component Analysis (ICA), Canonical Correlation Analysis (CCA), Minor Components Analysis (MCA)), Source Decomposition, Empirical Mode Decomposition (EMD), Support Vector Machine (SVM), and hybrid methods [3, 4, 29, 33–38]. A functional dedicated artifact channel which provides complementary aid to identify ECG/EOG is required to remove ocular or cardiac artifacts in the most of the available methods [4].

Regression is a common and well established technique in artifact removal, yet it cannot be used to remove muscle noise or line noise, since these type of artifacts have no reference channels [39]. Having a good regressor (e.g., an EOG) is critical in both time and frequency domain regression methods. It is an inherent weakness that eye movements and EEG signals are bidirectional. When unacceptable amount of data are lost in artifact rejection, delicate artifact removal methods which will preserve the essential EEG signals while removing artifacts are necessary [39]. One of the most important artifacts is EOG. EEG regions infected with EOG can be rejected from overall EEG signal with simplest artifact rejection where these portions are detected by EOG channels, however these regions still carry brain signals in addition to ocular artifacts and total rejection or subtraction of EOG from them results in loss of brain data [40–42].

Blind Source Separation (BSS) algorithms utilize multiple channels in an unsupervised learning algorithm to extract brain related activity from the ensemble EEG signal which can be assumed a linear superposition of brain signals, noise and artifacts [38]. Three common BSS algorithms are Independent Component Analysis (ICA), Principal Component Analysis (PCA) and Canonical Correlation Analysis (CCA).

ICA, a BSS method, is often used to remove EEG artifacts based on statistical approach of spatial filtering and separation of multiple channel EEG data into spatially fixed and temporally independent components [39, 43, 44]. Since the EEG sources and artifacts are usually of different origins, they can be assumed to be linear summation of each independent components. ICA method finds these statistically independent components and enable us to eliminate artifactual ones from the desired EEG [45]. On the other hand, ICA provides extraction of the eye related signals present in the EOG, and removal of this information or artifact, rather than the complete EOG which still has some brain activity [40], is possible. However, detection and removal of transient artifacts such as head and neck muscle contractions and movement are difficult with ICA [46]. Moreover, adapting ICA as an online method requires high computational power [46]. On the other hand, an advantage of ICA is that it does not rely on a reference channel [39]. However, many artifact removal algorithms are compared in [3], and Revised Aligned-Artifact Average (RAAA) and Second Order Blind Identification (SOBI) and Adaptive Mixture of Independent Component Analyzers (AMICA) are the preferred artifact removal methods for EOG, EMG and ECG artifacts.

PCA uses orthogonal transform of correlated time domain signal into linearly uncorrelated principal components (PCs) [47]. These principal components possess as much as variance of the EEG as possible. Artifact containing PCs can be eliminated if they are uncorrelated with the brain EEG. Application of PCA into ocular artifacts was provided in [48].

CCA is also another method utilized in removing artifacts. In CCA second order statistics are employed, correlation between two multivariate datasets are maximized by canonical variables. CCA offers shorter computational time compared to ICA [38].

Another method is filtering in frequency domain. Usually a high-pass filter starting from 0.5-1 Hz is applied for baseline drift removal. Notch filters are used to remove powerline-noise. Another one, EMG activity of contracting scalp sites can hinder the signals of interest in the EEG recordings during an epileptic seizure [49]. It was possible to remove this high frequency content EMG activity from EEG spectra by filtering out signals over 25 Hz. Adaptive Filters, Wiener Filtering and Bayesian Filters are three filtering methods applied in EEG signal preprocessing. Adaptive Filters are the most commonly used for artifact removal [47]. In Adaptive Filtering a reference channel for artifacts is subtracted from the EEG recursively. This reference is multiplied by a weight factor obtained from the output of the filter by a learning algorithm and this weighted reference is subtracted from the recorded EEG yielding output artifact free EEG changing adaptively [50].

In wavelet transform, many scaled and time shifted wavelets are used to produce coefficients for the particular signal and wavelet type by convolution of the signal and wavelets. These coefficients indicate similarity between the corresponding wavelet and the signal. In artifact removal via wavelet transform, the main idea is that the signal which can be highly correlated with a basis mother wavelet and can be separated from artifacts which might have no correlation to the principal mother wavelet [50]. Some examples of Wavelet Transform in artifact removal are for ocular artifact removal as in [51, 52].

3.3 EEG pre-processing pipelines available

Recently many preprocessing pipelines have been introduced in order to reduce the burden of artifact handling by an expert one by one visual inspection. This laborious task can be fastened by using existing automatized preprocessing methods in order. An efficient pre-processing pipeline not only helps the artifact management time but also provides objective evaluation with predefined criteria compared to highly subjective artifact handling by a human expert. The pre-processing pipelines usually consist of the combination of the following stages; filtering, re-referencing, bad channel identification (and interpolation), bad channel and epoch removal, artifact detection using ICA, artifact correction and removal [53], see **Figure 4**.

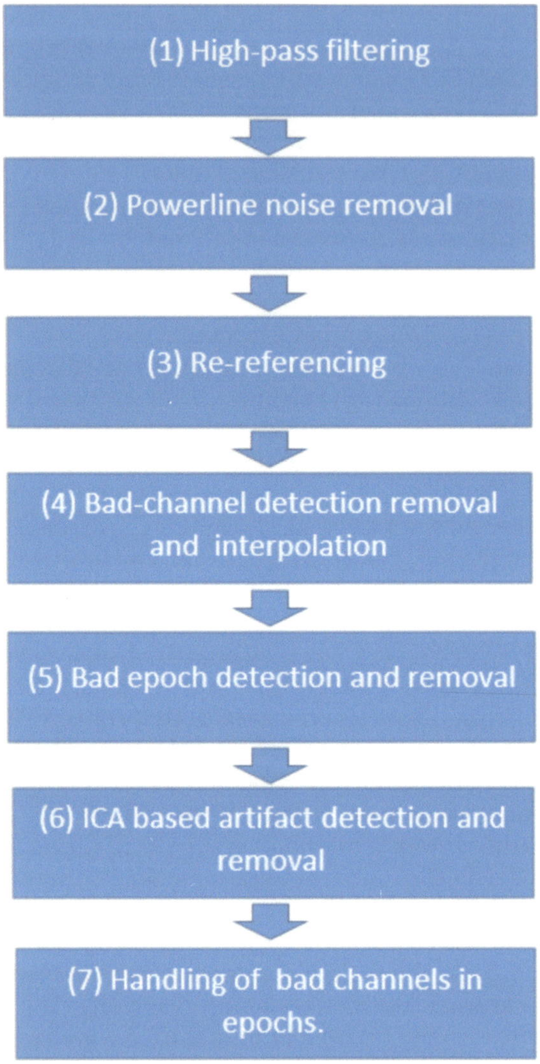

Figure 4.
APP artifact management flow diagram from [53].

Fully Automated Statistical Thresholding for EEG artifact Rejection (FASTER) [54] algorithm is a state of the art method which is available in EEGLAB toolbox [55]. FASTER has filtering, line noise removal, bad channel detection and interpolation, segmentation, and artifact rejection on segments by identifying bad channels, blinks, eye movements and muscular artifacts using combination of statistical thresholding and ICA [56]. It requires an extra EOG channel. The Automatic Pre-processing Pipeline (APP) removes powerline noise, bad channels, eye movements, blinks and muscular artifacts using ICA to identify artifactual components [53], see **Figure 4**. However, it also requires extra EOG channels. Da Cruz et al. [53] has found that APP performs better than FASTER yielding higher amplitude in ERP study. Another pipeline is Tool for Automated Processing of EEG data (TAPEEG) [57]. It uses automated routines of FASTER and Fieldtrip for artifact identification and performed similar to visually analysis by an expert [58]. TAPEEG handles the resting state EEG data as well. Both FASTER and TAPEEG are based on z- scores and have difficulty in handling outliers, this leads to loss of signal content due to false positive artifact detection and rejections [53]. Another standardized preprocessing method for large EEG datasets, PREP pipeline, handles line noise removal, bad channel detection, and referencing to standardize and normalize the data before processing [58]. It is also available as plug-in in EEGLAB toolbox.

Automagic is a toolbox developed for standardized handling of large growing EEG/ERP datasets by time [56]. The power of Automagic comes from the fact that it exploits many existing pipelines and methods, such as PREP pipeline for bad channel identification and for average referencing, Cleanline [59] to remove power line noise, EOG regression [60], Multiple Artifact Rejection Algorithm (MARA), ICA or robust PCA for artifact correction [61]. MARA is a plug-in available in EEGLAB which automatically identifies artifacts not only ocular or muscular but also any general artifactual source component in ICA [61]. Pedroni et al. [59] showed that combination of a preprocessing pipeline to identify bad channels and MARA method is efficient to remove most of the artifacts.

None of the methods offers a perfect robust and high accurate management of all types of artifacts. In general, they are all limited with the training dataset and fail to achieve high success with new type of artifactual data.

3.4 Simultaneous EEG and f-MRI artifact handling

Since EEG is widely used as a clinical tool to monitor or diagnose patients, doctors can be misguided in case of artifacts and EEG can be misinterpreted. For this reason, artifact removal becomes a crucial point for some cases such as epilepsy monitoring in an EEG/fMRI recording room. Today EEG and fMRI are two distinct but closely related and complementary methods. While fMRI provides high spatial resolution for localization of phenomena in the brain, EEG on the other hand results in better temporal resolution [62–65]. One should be careful about the experiments involving both fMRI and EEG because there are many unwanted electromagnetic sources interfering with EEG. For example, the false identification of spikes are highly possible since residuals of Ballistocardiogram (BCG) artifacts have similar shapes as epileptic spikes [66]. The factors that can lead to differences in the artifact are linked to the subject and experimental setup, [67]. There are imaging artifacts, cardiac related Ballistocardiogram artifacts (BCG), EOG and EMG artifacts in an EEG inside MRI [44]. Static field (B0) and the time-varying fields of radio-frequency excitations and of imaging gradients, generate artifacts in the EEG known as Ballistocardiogram (BCG) and imaging artifacts [44, 68–70]. The pulse artifact which can be observed in EEGs recorded inside MR scanners easily, is due to a fundamental cause that any movement of electrically conductive muscles in a static magnetic field generates electromagnetic induction and it is proportional to the static field, generally larger at higher field strengths [67, 71]. Pulsations of the scalp arteries are the main cause of this type of BCG artifact [72, 73]. The study of Grouiller et al. [44] compared different imaging artifact removal techniques and various cardiac artifact correction techniques in both simulated EEG data and in real experimental data. They concluded that there is no key for every door, some algorithms work well for some case and others might work well for other cases. Certain algorithms may be preferred depending on the type of data and analysis method [44]. Another algorithm, adaptive Optimal Basis Set (aOBS), automatically eliminates BCG artifacts yet preserving the neural origin signals in EEG [74]. It can be used efficiently for simultaneous fMRI and EEG recordings.

3.5 Sleep stage classification artifact handling

Manual artifact detection is still the most common method for artifact handling for sleep stage classification, however, the long time required and the difficulty to apply it to large datasets poses the main disadvantages [75]. Malafeev et al. [75] compared 12 simple algorithms that are applicable with a single EEG channel for ease of use. It was found that automatic artifact detection in EEG during sleep within large datasets is possible with simple algorithms. Among these, Power thresholding 25–90 Hz (PT25), Power thresholding 45–90 Hz (PT45) and Autoregressive (AR) models had Reciever Operating Characteristic (ROC) areas above 0.95. In addition, online detection is also possible with the majority of these simple algorithms.

3.6 BCI Artifact handling

Artifact removal in BCI applications are getting more attention. By studies it was shown that artifacts generated by EOG and EMG activities affect the neurological signals utilized in a BCI system [10, 76]. Although there are extensive researches into artifact removal for BCIs and developed efficient methods such as Fully Online and Automated Artifact Removal (FORCe), Lagged Auto-Manual Information Clustering (LAMIC), Fully Automated Statistical Thresholding for EEG artifact Rejection (FASTER) and K-Singular Value Decomposition (K-SVD), the field lacks an effective artifact removal [12, 54, 77–82]. The surrogate-based artifact removal (SuBAR) technique proposed by Chavez et al. [33] effectively cancels EOG and EMG artifacts from single-channel EEG. Chang et al. [83] proposed a method for detection of eye artifact from single prefrontal channel which is useful for headband-type wearable EEG devices with a few frontal EEG channels. Compared to conventional methods the accuracy of detecting ocular artifact contaminated epochs was significantly better. Daily-life EEG-BCIs are getting popular and artifact removal techniques for these BCIs must have some critical features such as; must be performed outdoor, with portable wearable wireless device, with real EEG signals, compatible with daily life tasks, must have simple electrical montage, must use dry electrodes, must remove complex artifacts, must work only EEG without reference, must work online and must work with single electrode channel. More research into artifact removal other than ocular and cardiac artifacts is necessary especially for those daily-life EEG BCIs [36].

While ICA and PCA are common artifact removal methods, Artifact Subspace Reconstruction (ASR), which is a powerful automated artifact removal method available for both online real-time and offline, can be applied to prevent transient and large artifact [46, 84]. It also does not require additional channel and cleans the data from artifacts.

4. Conclusion

The number of artifact handling techniques and algorithms are increasing drastically, however the artifact problem is still challenging for many applications. Particularly, the internal or physiologic artifacts are difficult to distinguish and remove. While simple measures such as artifact avoidance and artifact rejection can be utilized in some applications, most of the cases require special methods dedicated to handle artifacts in order to significantly reduce their harmful effects on signal of interest. Due to the varying nature of artifacts a generic method for all sorts of artifacts is still missing. However preprocessing pipelines provides some efficient approaches to this challenge. In future, the progress in machine learning and deep learning based approaches may yield more efficient, accurate and robust artifact removal options. Online artifact removal methods such as ASR must be developed to overcome various artifacts in daily life to be efficient for BCIs.

Author details

İbrahim Kaya
Department of Biomedical Engineering, Izmir Katip Celebi University, Izmir, Turkey

*Address all correspondence to: ibrahimkaya21@yahoo.com

References

[1] Priemer R. Introductory signal processing. Vol. 6. World Scientific; 1991.

[2] Da Silva FL. EEG: origin and measurement. InEEg-fMRI 2009 (pp. 19-38). Springer, Berlin, Heidelberg.

[3] Urigüen JA, Garcia-Zapirain B. EEG artifact removal—state-of-the-art and guidelines. Journal of neural engineering. 2015 Apr 2;12(3):031001.

[4] Islam MK, Rastegarnia A, Yang Z. Methods for artifact detection and removal from scalp EEG: A review. Neurophysiologie Clinique/Clinical Neurophysiology. 2016 Nov 1;46(4-5): 287-305.

[5] Sazgar M, Young MG. Absolute epilepsy and EEG rotation review. Spinger; 2019.

[6] Croft RJ, Barry RJ. Removal of ocular artifact from the EEG: a review. Neurophysiologie Clinique/Clinical Neurophysiology. 2000 Feb 1;30(1):5-19.

[7] Vigon L, Saatchi MR, Mayhew JE, Fernandes R. Quantitative evaluation of techniques for ocular artefact filtering of EEG waveforms. IEE Proceedings-Science, Measurement and Technology. 2000 Sep 1;147(5):219-28.

[8] Vigon L, Saatchi MR, Mayhew JE, Fernandes R. Quantitative evaluation of techniques for ocular artefact filtering of EEG waveforms. IEE Proceedings-Science, Measurement and Technology. 2000 Sep 1;147(5):219-28.

[9] Matsuo F, Peters JF, Reilly EL. Electrical phenomena associated with movements of the eyelid. Electro-encephalography and clinical neuro-physiology. 1975 May 1;38(5):507-11.

[10] Goncharova II, McFarland DJ, Vaughan TM, Wolpaw JR. EMG contamination of EEG: spectral and topographical characteristics. Clinical neurophysiology. 2003 Sep 1;114(9):1580-93.

[11] McMenamin BW, Shackman AJ, Maxwell JS, Bachhuber DR, Koppenhaver AM, Greischar LL, Davidson RJ. Validation of ICA-based myogenic artifact correction for scalp and source-localized EEG. Neuroimage. 2010 Feb 1;49(3):2416-32.

[12] Sweeney KT, Ayaz H, Ward TE, Izzetoglu M, McLoone SF, Onaral B. A methodology for validating artifact removal techniques for physiological signals. IEEE transactions on information technology in biomedicine. 2012 Jul 10;16(5):918-26.

[13] Anderer P, Roberts S, Schlögl A, Gruber G, Klösch G, Herrmann W, Rappelsberger P, Filz O, Barbanoj MJ, Dorffner G, Saletu B. Artifact processing in computerized analysis of sleep EEG–a review. Neuropsychobiology. 1999; 40(3):150-7.

[14] McFarland DJ, McCane LM, David SV, Wolpaw JR. Spatial filter selection for EEG-based communication. Electroencephalography and clinical Neurophysiology. 1997 Sep 1;103(3): 386-94.

[15] Tamburro G, Stone DB, Comani S. Automatic Removal of Cardiac Interference (ARCI): a new approach for EEG data. Frontiers in neuroscience. 2019 May 8;13:441.

[16] Luck, S. J. Event-related potentials. In H. Cooper, P. M. Camic, D. L. Long, A. T. Panter, D. Rindskopf, & K. J. Sher (Eds.), (2012). APA handbooks in psychology®. APA handbook of research methods in psychology, Vol. 1. Foundations, planning, measures, and psychometrics (p. 523-546). American Psychological Association. https://doi.org/10.1037/13619-028

[17] Dawson GD. A summation technique for the detection of small evoked potentials. Electroencephalography & clinical neurophysiology. 1954.

[18] Hoke M, Ross B, Wickesberg R, Lütkenhöner B. Weighted averaging—theory and application to electric response audiometry. Electroencephalography and clinical neurophysiology. 1984 May 1;57(5):484-9.

[19] Bezerianos A, Laskaris N, Fotopoulos S, Papathanasopoulos P. Data dependent weighted averages for recording of evoked potential signals. Electroencephalography and Clinical Neurophysiology/Evoked Potentials Section. 1995 Sep 1;96(5):468-71.

[20] Davila, C. E., & Mobin, M. S. (1992). Weighted averaging of evoked potentials. IEEE Transactions on Biomedical Engineering, 39(4), 338-345

[21] John MS, Dimitrijevic A, Picton TW. Weighted averaging of steady-state responses. Clinical Neurophysiology. 2001 Mar 1;112(3):555-62.

[22] Lütkenhöner B, Hoke M, Pantev C. Possibilities and limitations of weighted averaging. Biological cybernetics. 1985 Oct;52(6):409-16.

[23] Mühler R, Specht HV. Sorted averaging-principle and application to auditory brainstem responses. Scandinavian audiology. 1999 Jan 1;28(3):145-9.

[24] Rahne T, von Specht H, Mühler R. Sorted averaging—application to auditory event-related responses. Journal of neuroscience methods. 2008 Jul 15;172(1):74-8.

[25] Yabe H, Saito F, Fukushima Y. Median method for detecting endogenous event-related brain potentials. Electroencephalography and clinical Neurophysiology. 1993 Dec 1;87(6):403-7.

[26] Özdamar Ö, Kalayci T. Median averaging of auditory brain stem responses. Ear and hearing. 1999 Jun 1;20(3):253-64.

[27] Leonowicz Z, Karvanen J, Shishkin SL. Trimmed estimators for robust averaging of event-related potentials. Journal of neuroscience methods. 2005 Mar 15;142(1):17-26.

[28] Leski JM, Gacek A. Computationally effective algorithm for robust weighted averaging. IEEE transactions on biomedical engineering. 2004 Jun 21;51(7):1280-4.

[29] Fatourechi M, Bashashati A, Ward RK, Birch GE. EMG and EOG artifacts in brain computer interface systems: A survey. Clinical neurophysiology. 2007 Mar 1;118(3):480-94.

[30] Ochoa CJ, Polich J. P300 and blink instructions. Clinical Neurophysiology. 2000 Jan 1;111(1):93-8.

[31] Verleger R. The instruction to refrain from blinking affects auditory P3 and N1 amplitudes. Electroencephalography and Clinical Neurophysiology. 1991 Mar 1;78(3):240-51.

[32] Hagemann D, Naumann E. The effects of ocular artifacts on (lateralized) broadband power in the EEG. Clinical Neurophysiology. 2001 Feb 1;112(2):215-31.

[33] Chavez M, Grosselin F, Bussalb A, Fallani FD, Navarro-Sune X. Surrogate-based artifact removal from single-channel EEG. IEEE transactions on neural systems and rehabilitation engineering. 2018 Jan 22;26(3):540-50.

[34] Chen Y, Zhao Q, Hu B, Li J, Jiang H, Lin W, Li Y, Zhou S, Peng H. A method of removing ocular artifacts from EEG using discrete wavelet transform and Kalman filtering. In2016 IEEE International Conference on Bioinformatics and Biomedicine (BIBM) 2016 Dec 15 (pp. 1485-1492). IEEE.

[35] Lins OG, Picton TW, Berg P, Scherg M. Ocular artifacts in recording EEGs and event-related potentials II: Source dipoles and source components. Brain topography. 1993 Sep;6(1):65-78.

[36] Minguillon J, Lopez-Gordo MA, Pelayo F. Trends in EEG-BCI for daily-life: Requirements for artifact removal. Biomedical Signal Processing and Control. 2017 Jan 1;31:407-18.

[37] Shao SY, Shen KQ, Ong CJ, Wilder-Smith EP, Li XP. Automatic EEG artifact removal: a weighted support vector machine approach with error correction. IEEE Transactions on Biomedical Engineering. 2008 Oct 3;56(2):336-44.

[38] Jiang X, Bian GB, Tian Z. Removal of artifacts from EEG signals: a review. Sensors. 2019 Jan;19(5):987.

[39] Jung TP, Humphries C, Lee TW, Makeig S, McKeown MJ, Iragui V, Sejnowski TJ. Extended ICA removes artifacts from electroencephalographic recordings. Advances in neural information processing systems. 1998 Nov 30:894-900.

[40] Vigário RN. Extraction of ocular artefacts from EEG using independent component analysis. Electroencephalography and clinical neurophysiology. 1997 Sep 1;103(3):395-404.

[41] Barlow JS. Computerized clinical electroencephalography in perspective. IEEE Transactions on Biomedical Engineering. 1979 Jul(7):377-91.

[42] Verleger R. Valid identification of blink artefacts: are they larger than 50 μV in EEG records?. Electroencephalography and clinical Neurophysiology. 1993 Dec 1;87(6):354-63.

[43] Jung TP, Makeig S, Humphries C, Lee TW, Mckeown MJ, Iragui V, Sejnowski TJ. Removing electroencephalographic artifacts by blind source separation. Psychophysiology. 2000 Mar;37(2):163-78.

[44] Grouiller F, Vercueil L, Krainik A, Segebarth C, Kahane P, David O. A comparative study of different artefact removal algorithms for EEG signals acquired during functional MRI. Neuroimage. 2007 Oct 15;38(1):124-37.

[45] Makeig S, Bell AJ, Jung TP, Sejnowski TJ. Independent component analysis of electroencephalographic data. Advances in neural information processing systems. 1996 Dec 2:145-51.

[46] Chang CY, Hsu SH, Pion-Tonachini L, Jung TP. Evaluation of artifact subspace reconstruction for automatic EEG artifact removal. In2018 40th Annual International Conference of the IEEE Engineering in Medicine and Biology Society (EMBC) 2018 Jul 18 (pp. 1242-1245). IEEE.

[47] Mannan MM, Kamran MA, Jeong MY. Identification and removal of physiological artifacts from electroencephalogram signals: A review. Ieee Access. 2018 May 31;6:30630-52.

[48] Berg P, Scherg M. Dipole modelling of eye activity and its application to the removal of eye artefacts from the EEG and MEG. Clinical Physics and Physiological Measurement. 1991;12(A):49.

[49] Gotman J, Ives JR, Gloor P. Frequency content of EEG and EMG at seizure onset: possibility of removal of EMG artefact by digital filtering. Electroencephalography and clinical neurophysiology. 1981 Dec 1;52(6):626-39.

[50] Kim SP. Preprocessing of eeg. InComputational EEG Analysis 2018 (pp. 15-33). Springer, Singapore.

[51] Krishnaveni V, Jayaraman S, Anitha L, Ramadoss K. Removal of ocular artifacts from EEG using adaptive thresholding of wavelet

coefficients. Journal of neural engineering. 2006 Nov 23;3(4):338.

[52] Zikov T, Bibian S, Dumont GA, Huzmezan M, Ries CR. A wavelet based de-noising technique for ocular artifact correction of the electroencephalogram. InProceedings of the Second Joint 24th Annual Conference and the Annual Fall Meeting of the Biomedical Engineering Society] [Engineering in Medicine and Biology 2002 Oct 23 (Vol. 1, pp. 98-105). IEEE.

[53] da Cruz JR, Chicherov V, Herzog MH, Figueiredo P. An automatic pre-processing pipeline for EEG analysis (APP) based on robust statistics. Clinical Neurophysiology. 2018 Jul 1; 129(7):1427-37.

[54] Nolan H, Whelan R, Reilly RB. FASTER: fully automated statistical thresholding for EEG artifact rejection. Journal of neuroscience methods. 2010 Sep 30;192(1):152-62.

[55] Delorme A, Makeig S. EEGLAB: an open source toolbox for analysis of single-trial EEG dynamics including independent component analysis. Journal of neuroscience methods. 2004 Mar 15;134(1):9-21.

[56] Pedroni A, Bahreini A, Langer N. Automagic: Standardized preprocessing of big EEG data. NeuroImage. 2019 Oct 15;200:460-73.

[57] Hatz F, Hardmeier M, Bousleiman H, Rüegg S, Schindler C, Fuhr P. Reliability of fully automated versus visually controlled pre-and post-processing of resting-state EEG. Clinical Neurophysiology. 2015 Feb 1;126(2):268-74.

[58] Bigdely-Shamlo N, Mullen T, Kothe C, Su KM, Robbins KA. The PREP pipeline: standardized preprocessing for large-scale EEG analysis. Frontiers in neuroinformatics. 2015 Jun 18;9:16.

[59] Mullen T. CleanLine EEGLAB plugin. San Diego, CA: Neuroimaging Informatics Toolsand Resources Clearinghouse (NITRC). 2012.

[60] Parra LC, Spence CD, Gerson AD, Sajda P. Recipes for the linear analysis of EEG. Neuroimage. 2005 Nov 1;28(2): 326-41.

[61] Winkler I, Haufe S, Tangermann M. Automatic classification of artifactual ICA-components for artifact removal in EEG signals. Behavioral and Brain Functions. 2011 Dec;7(1):1-5.

[62] Huster RJ, Debener S, Eichele T, Herrmann CS. Methods for simultaneous EEG-fMRI: an introductory review. Journal of Neuroscience. 2012 May 2;32(18):6053-60.

[63] Vanni S, Warnking J, Dojat M, Delon-Martin C, Bullier J, Segebarth C. Sequence of pattern onset responses in the human visual areas: an fMRI constrained VEP source analysis. Neuroimage. 2004 Mar 1;21(3):801-17.

[64] Wibral M, Bledowski C, Kohler A, Singer W, Muckli L. The timing of feedback to early visual cortex in the perception of long-range apparent motion. Cerebral cortex. 2009 Jul 1; 19(7):1567-82.

[65] Wibral M, Bledowski C, Turi G. Integration of separately recorded EEG/ MEG and fMRI data. Simultaneous EEG and fMRI: recording, analysis, and application (Ullsperger M, Debener S, eds). 2010 May 28:209-34.

[66] de Munck JC, van Houdt PJ, Gonçalves SI, van Wegen E, Ossenblok PP. Novel artefact removal algorithms for co-registered EEG/fMRI based on selective averaging and subtraction. Neuroimage. 2013 Jan 1;64:407-15.

[67] Debener S, Kranczioch C, Gutberlet I. EEG quality: origin and reduction of the EEG cardiac-related artefact. InEEG-fMRI 2009 (pp. 135-151). Springer, Berlin, Heidelberg.

[68] Bonmassar G, Purdon PL, Jääskeläinen IP, Chiappa K, Solo V, Brown EN, Belliveau JW. Motion and ballistocardiogram artifact removal for interleaved recording of EEG and EPs during MRI. Neuroimage. 2002 Aug 1;16(4):1127-41.

[69] Allen PJ, Josephs O, Turner R. A method for removing imaging artifact from continuous EEG recorded during functional MRI. Neuroimage. 2000 Aug 1;12(2):230-9.

[70] Felblinger J, Slotboom J, Kreis R, Jung B, Boesch C. Restoration of electrophysiological signals distorted by inductive effects of magnetic field gradients during MR sequences. Magnetic Resonance in Medicine: An Official Journal of the International Society for Magnetic Resonance in Medicine. 1999 Apr;41(4):715-21.

[71] Debener S, Mullinger KJ, Niazy RK, Bowtell RW. Properties of the ballistocardiogram artefact as revealed by EEG recordings at 1.5, 3 and 7 T static magnetic field strength. International Journal of Psychophysiology. 2008 Mar 1;67(3):189-99.

[72] Allen PJ, Polizzi G, Krakow K, Fish DR, Lemieux L. Identification of EEG events in the MR scanner: the problem of pulse artifact and a method for its subtraction. Neuroimage. 1998 Oct 1;8(3):229-39.

[73] Ives JR, Warach S, Schmitt F, Edelman RR, Schomer DL. Monitoring the patient's EEG during echo planar MRI. Electroencephalography and clinical neurophysiology. 1993 Dec 1;87(6):417-20.

[74] Marino M, Liu Q, Koudelka V, Porcaro C, Hlinka J, Wenderoth N, Mantini D. Adaptive optimal basis set for BCG artifact removal in simultaneous EEG-fMRI. Scientific reports. 2018 Jun 11;8(1):1-1.

[75] Malafeev A, Omlin X, Wierzbicka A, Wichniak A, Jernajczyk W, Riener R, Achermann P. Automatic artefact detection in single-channel sleep EEG recordings. Journal of sleep research. 2019 Apr;28(2):e12679.

[76] McFarland DJ, Sarnacki WA, Vaughan TM, Wolpaw JR. Brain-computer interface (BCI) operation: signal and noise during early training sessions. Clinical Neurophysiology. 2005 Jan 1;116(1):56-62.

[77] Chen X, Liu A, Peng H, Ward RK. A preliminary study of muscular artifact cancellation in single-channel EEG. Sensors. 2014 Oct;14(10):18370-89.

[78] Chen X, Liu A, Chiang J, Wang ZJ, McKeown MJ, Ward RK. Removing muscle artifacts from EEG data: Multichannel or single-channel techniques?. IEEE Sensors Journal. 2015 Dec 8;16(7):1986-97.

[79] Daly I, Nicolaou N, Nasuto SJ, Warwick K. Automated artifact removal from the electroencephalogram: a comparative study. Clinical EEG and neuroscience. 2013 Oct;44(4):291-306.

[80] Daly I, Scherer R, Billinger M, Müller-Putz G. FORCe: Fully online and automated artifact removal for brain-computer interfacing. IEEE transactions on neural systems and rehabilitation engineering. 2014 Aug 13;23(5):725-36.

[81] Khatun S, Mahajan R, Morshed BI. Comparative study of wavelet-based unsupervised ocular artifact removal techniques for single-channel EEG data. IEEE journal of translational engineering in health and medicine. 2016 Mar 22; 4:1-8.

[82] Sreeja SR, Sahay RR, Samanta D, Mitra P. Removal of eye blink artifacts from EEG signals using sparsity. IEEE journal of biomedical and health informatics. 2017 Nov 13;22(5):1362-72.

[83] Chang WD, Lim JH, Im CH. An unsupervised eye blink artifact detection method for real-time electroencephalogram processing. Physiological measurement. 2016 Feb 19;37(3):401.

[84] Kothe CA, Jung TP, inventors. Artifact removal techniques with signal reconstruction. United States patent application US 14/895,440. 2016 Apr 28.

Section 2

Therapeutic Brain-Computer Interfacing

Chapter 3

Pain Identification in Electroencephalography Signal Using Fuzzy Inference System

Vahid Asadpour, Reza Fazel-Rezai, Maryam Vatankhah and Mohammad-Reza Akbarzadeh-Totonchi

Abstract

Diagnosing pain mechanisms is one of main approaches to improve clinical treatments. Especially, detection of existence and/or level of pain could be vital when oral information is not present for instant for neonates, disabled persons, anesthetized patients and animals. Various researches have been performed to originate and classify the pain; however, consistent results are surprising. The aim of this study is to show a strict relation between electroencephalography (EEG) features and perceptual pain levels and to clarify the relation of classified signal to pain origin. Cortical regions on scalp are assigned based on an evolutional method for optimized alignment of electrodes that improve the clinical monitoring results. The EEG signals are recorded during relax condition and variety of pain conditions. Evolutionary optimization method is used to reduce the features space dimension and computational costs. A hybrid adaptive network fuzzy inference system (ANFIS) and support vector machine (SVM) scheme is used for classification of pain levels. ANFIS optimizer is used to fine tune the non-linear alignment of kernels of SVM. The results show that pain levels could be differentiated with high accuracy and robustness even for few recording electrodes. The proposed classification method provides up to 95% accuracy.

Keywords: electro-encephalogram, pain, adaptive network fuzzy inference system, support vector machine

1. Introduction

Diagnosis of the pain is one of the main concerns in clinical treatments procedure. In particular, detection of chronic or acute stage of the pain could be vital in the situations that oral information is not available for example instant for neonates, disabled persons, anesthetized patients and also animals. Multiple research projects have been done to originate and classify the pain. It is shown that achieving consistent is a challenge.

Identification of the human sensory perception have been of high interest in recent years. These studies are required for protection of the body and for restoring the embodiment sense. The advance in this field shows that not only the accurate design of the sensors improves the sensitivity of the identification but understanding the dynamics of pain perception and successful reversing of the coding mechanism are essential stages of the processing and classification process.

Localization of the source of the pain is very important for the neurological therapeutic processes [1]. The localization of cortical sources and observation of the spatiotemporal activation is also used for pre-treatment monitoring and surgical process [2]. The studies in this area would create an infrastructure for real-time monitoring of the pain to be used in alarming systems, surgery monitors and automated activated systems.

The aim of this work is to show the relation of EEG signal and perceptual level of pain. We also try to clarify the relation between the signal and the origin of the pain. The alignment of electrodes in cortical regions on scalp are assigned based on an evolutional algorithm to improve the clinical monitoring results. The normal and pain conditions are used for recording the signal. Some defined spectral features are combined with non-linear features including approximate entropy and Lyapunov exponent to create the feature vector. It is shown that there is consistency between these features and the dynamical characteristic of EEG signals. Evolutionary optimization method is used for reduction of the features space dimension and computational costs. A hybrid adaptive network fuzzy inference system (ANFIS) and support vector machine (SVM) scheme is used as the classifier. ANFIS optimizer is used for alignment of kernels of SVM. The classification results show that pain levels could be differentiated with high accuracy, sensitivity, and specificity with few recording electrodes. This research shows that electrical variations of brain patterns could be used for determination of pain levels. The proposed classification method reaches an accuracy of 95%.

2. Literature review

The study of human brain functionality in special conditions like stress and pain has significantly improved in the last decades [2]. Only some few changes in EEG signal have been observed during pain condition. An experimental pain stimulus will cause a decrease in alpha spectrum and an increase in gamma power in the surface of cortex while tonic muscle pain usually led to a stronger beta activity [3]. In most of the works a cold press has been mainly used to induce pain to the subjects. The achieved results are not necessarily consistent and do not allow for generalization of the events.

The staging of the signals has shown more progress regardless of the specificity of the described EEG changes for pain. The EEG spectrum is affected by sensory processing in general and cognitive sensory signals change during these events. The ambiguity of the effects of these events on EEG is probably a consequence of the methods used for EEG analysis, which do not allow for sufficient experimental control. In the present work a machine learning approach is used for classification and recognition of pain to be used for diagnosing purposes.

The aim of this research is to identify the difference between "normal", "low pain" and "pain" conditions. A kernel based SVM is used for the classification of the signals in the desired classes [4]. The optimized hyperplane is adjusted by finding the maximum distance from the nearest training points. An ANFIS optimizer is used for adjusting the hyper-planes of SVM classifier. ANFIS is trained by the features in the data set and adjusts the system parameters according to the error criteria [5]. Our results show that the combination of ANFIS-SVM results to the best performance on nonlinear features.

3. Materials and methods

3.1 Energy ratio features

The EEG signal could be categorized based on the spectrum in the frequency domain. The spectrum analysis could provide a demonstration of the functionality of the brain. Because of the spectrum changes in pain condition the Energy ratios between different bands could be used as the classification features. The ratio of Alpha, Beta, Delta, and Theta energy to the total spectrum on each EEG lead are used as the features for the classifier.

3.2 Approximate entropy

Approximate entropy is a non-negative number that is assigned to a time series that is a measure of the complexity or irregularity in the data. EEG signal has a steady pattern during synchronized cooperative function of cortical cells with low entropy index values. In contrast concentric functions and higher levels of brain activity led to high values of entropy. The entropy H is defined as:

$$H = -\sum_{i=1}^{N} P_i \log_2 P_i \tag{1}$$

in which P_i is the average probability at i th frequency band of brain rhythm that is grater that r times of standard deviation and N is the total number of frequency bands. H is 0 for a single frequency and 1 for uniform spectrum distribution over total spectrum. Approximate entropy can be used as a powerful tool in the study of the EEG activity because of the non-linear characteristics of EEG signals. The accuracy and confidence of the entropy estimate improves as the number of matches of length m and $m + 1$ increases. m and r are critical in finding the outcome of approximate entropy. The approximate entropy is estimated with $m = 3$ and $r = 0.25$ based on an investigation on original data sequence in this work.

3.3 Fractal dimension

Fractal dimension is a demonstration of the geometric property of basin of attraction in the feature space. This dimension shows geometrical property of attractors and is also computed very fast [6]. Features were extracted from each one second segment with 50% overlap, and sequence of 9 extracted features was considered as the feature vector of a five second segment. We have used Higuchi's algorithm, in which k new time series are constructed from the signal under study as [7]:

$$x_m^k = \left\{ x(m), x(m+k), x(m+2k), \ldots, x\left(m + \lfloor \frac{N-m}{k} \rfloor k \right\} \tag{2}$$

in which $m = 1, 2, \ldots, k$ and k indicate the initial time value, and the discrete time interval between points, respectively. For each of the k time series x_m^k the length $L_m(k)$ is calculated as:

$$L_m(k) = \frac{\sum_{i=1} |x(m+ik) - x(m+(i-1)k|(N-1)}{\lfloor \frac{N-m}{k} \rfloor k} \tag{3}$$

in which N is the total length of the signal x. An average length is computed as the mean of the k lengths $L_m(k)$ (for $m = 1, 2, \ldots, k$). This procedure is repeated for each k ranging from 1 to k_{max} , obtaining an average length for each k. In the curve of $ln\ (L(k))$ versus $ln\ \left(\frac{1}{k}\right)$, the slope of the best matched line to this curve is the estimate of the fractal dimension.

3.4 Lyapunov exponent

Lyapunov exponents are used as a measure for differentiating between types of orbits in feature space based on the initial conditions. These features can determine the stability of steady-state and chaotic behavior [8]. Chaotic systems show aperiodic dynamics because the phase space trajectories with similar initial states tend to move from each other at an exponentially increasing speed that is defined as Lyapunov exponent. This feature is extracted from the observed time series. The algorithm starts from the two nearest neighboring points in phase space at the beginning time 0 and at the current time t that corresponds to the distances of the points in the i th direction are $\|\delta X_i(0)\|$ and $\|\delta X_i(t)\|$, respectively. The Lyapunov exponent is defined as the average growth rate λ_i of the initial distance [9]:

$$\frac{\|\delta X_i(t)\|}{\|\delta X_i(0)\|} = 2^{\lambda_i}(\mathrm{t} \to \infty) \tag{4}$$

or

$$\lambda_i = \lim_{t \to \infty} \frac{1}{t} \log_2 \frac{\|\delta X_i(t)\|}{\|\delta X_i(0)\|} \tag{5}$$

The existence of a positive Lyapunov exponent is an indication of chaos. Lyapunov exponents can be extracted from observed signals using two approaches. The first method is based on the following of the time-evolution of nearby points in the state space. This method can only estimate the largest Lyapunov. In the other approach the Jacobi matrices and can estimate all the Lyapunov exponents for a systems that often called their Lyapunov spectra [10]. This vector is used as the parameter vector in this work.

4. Classification

SVM classifiers discriminants the hyperplanes to reach optimal classification. The hyperplanes should be adjusted to maximize the margin of classification boundaries. The distance from the nearest training points is measured using a non-linear kernel to map the problem from the feature space into the linear space [11]. Radial Basis Function (RBF) kernel is proposed in this paper and the Lagrangian optimization of the kernel is performed using an ANFIS structure. This proposed method leads to adjustable soft-decision classification because of the conceptual nature of the pain for patients.

4.1 SVM with RBF kernel

Training of the SVM is a quadratic optimization problem on the hyperplanes that are defined as:

$$y_i\left(w\Phi\left(x_i, y_j\right) + b\right) \geq 1 - \xi_i, \ \ \xi_i \geq 0, \ \ i = 1, \dots, l, \ \ j = 1, \dots, m \tag{6}$$

that x_i is the feature vector, b is the bias, w is the weight vector, ξ_i is class separation factor, $\Phi\left(x_i, y_j\right)$ is the RBF mapping kernel, l is the number of training vectors, j is the number of output vectors, and y_j is the desired output vector. The weight parameter should be optimized to maximize the margin between the hyperplane and the neighboring points in the feature space. This is a compromise between the maximization of margin and the number of misclassified points. Optimization of Eq. (6) results to optimum weight w.

Figure 1 shows the RBF kernel SVM classification system. The kernel parameters could be selected by optimizing the upper bound of the generalization error based on the training data. The support vector fractions and the relation between the number of support vectors and all the training samples define an upper bound on the error estimate. The resulting decision function can only change when support vectors are excluded. A low fraction of support vectors could be used as a criterion for the parameter selection.

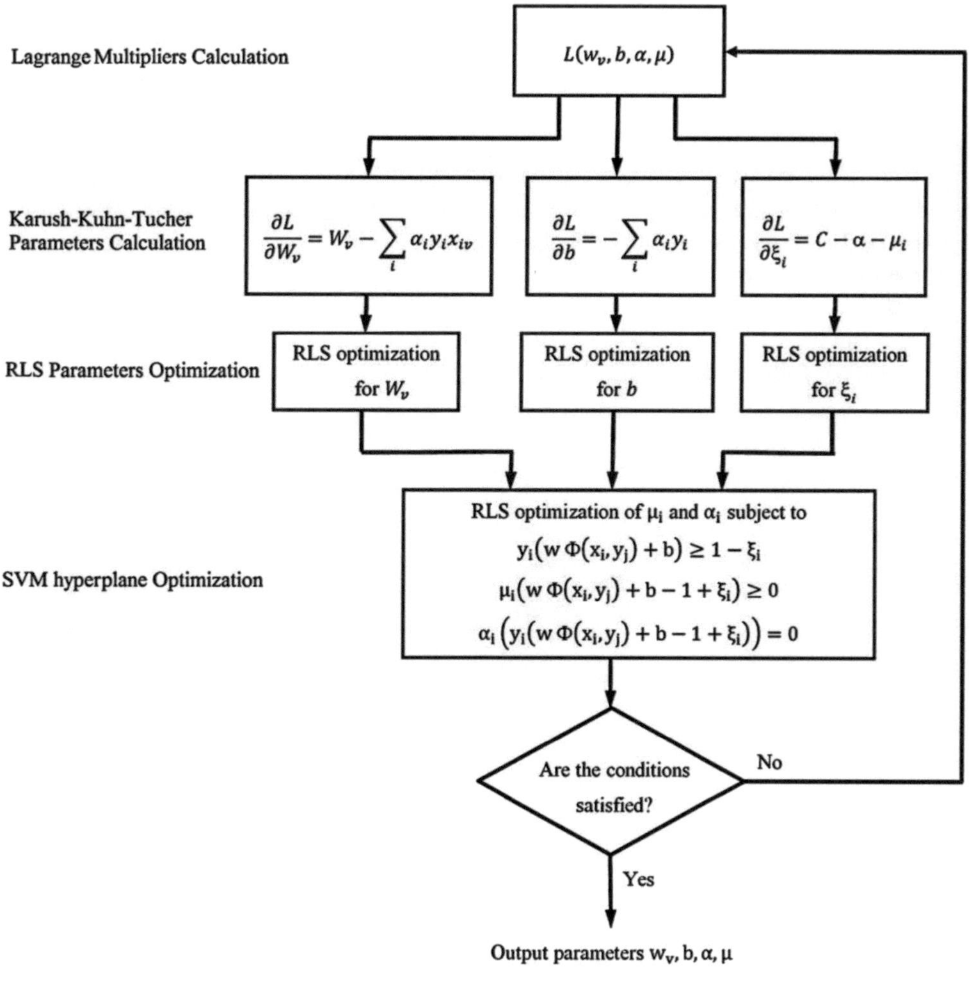

Figure 1.
Block diagram of RBF-SVM classification system.

4.2 Adjustable ANFIS optimization

An ANFIS is used with for optimization of the SVM classification kernels. The optimization process would be less reliant on expert knowledge compared to the conventional fuzzy systems using this adaptive method. The result of the learning algorithm for this architecture is to adjust all the parameters of the kernel to adjust the hyperplanes for optimized output. Since the initial parameters are not fixed, the search space becomes larger, and the convergence of the training becomes slower. The training method is using a combining of the least squares and the gradient descent method is used to train the network. The hybrid algorithm is composed of forward and backward pass. The least squares method on the forward pass is used to optimize the consequent parameters with the fixed premise parameters. The backward pass is using the gradient descent method afterward to optimize the consequent parameters and to adjust the premise parameters corresponding to the fuzzy sets in the input domain. The output of the ANFIS network is achieved by defuzzification of the consequent parameters in the forward pass. The output error is used to adjust the premise parameters using a standard back propagation algorithm.

5. Results

Three montages of electrodes are used for clinical experiments as are shown in **Figure 2**. The classification results of for the arrangements are compared with each other to find the best montage. All of the experiments performed with 70% training and 30% testing signals. **Table 1** shows the classification accuracy rates for ANFIS-SVM using these electrode arrangements. Montage III led to best classification and is used as the electrode set for pain classification in this work. The SVM parameters and the related Kernel function are adjusted to achieve the best possible results. This optimization was done using ANFIS as described in the materials and methods section. The over fitting is reduced by controlling the compromise between the training error minimization and the learning capacity of the fuzzy if-then functions.

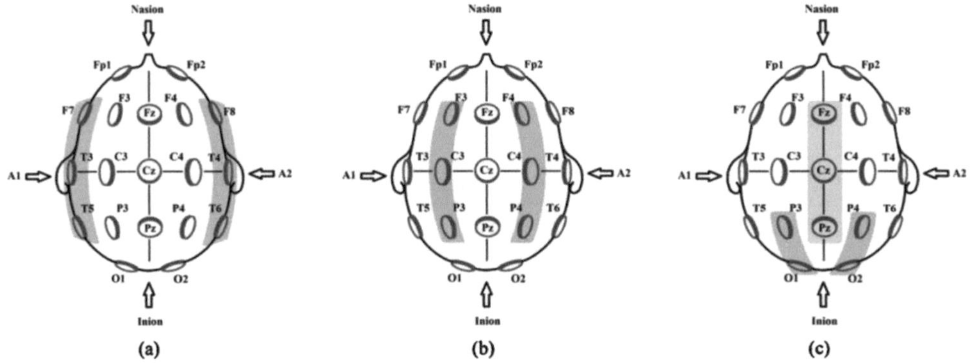

Figure 2.
The electrode arrangement (a) I, (b) II, and (c) III for electrodes based on 10/20 standard.

	Montage I	Montage II	Montage III
Classification rate for ANFIS-SVM	78%	81%	88%

Table 1.
ANFIS-SVM Classification rate for three electrode arrangements.

	Features	SVM (%)	ANFIS-SVM (%)
Non-reduced features	Standard deviation, theta ratio, alpha ratio, entropy, lyapunov, and fractal dimension	90	95
Reduced Features	Theta ratio, alpha ratio, and entropy	83	87

Table 2.
Classification rates of SVM and ANFIS-SVM for reduced and non-reduced features.

	Spectral features (%)	Non-linear features (%)
SVM	75	89
ANFIS-SVM	80	93

Table 3.
Classification rate for ANFIS-SVM with two sets of features.

The final decision function parameters can be updated because they depend on the support vectors only.

Furthermore, approximate entropy, Lyapunov exponent and fractal dimension are also examined as non-linear features. An evolutionary feature selection was applied on these elements that showed the theta and alpha ratio and the entropy led to best classification rates for ANFIS-SVM classifier. The accuracies for classification of two classes of pain and no pain are shown in **Table 2** for two cases of using all the features in the feature vector and using the high rank features only. The results for SVM and ANFIS-SVM show the identification rate shows a reduction of 7% and 8% for reduced features, respectively. This reduction happens for identification based on three most effective features, and it could be concluded that the effect of "standard deviation" and "fractal dimension" could not be neglected. The accuracy of 95 % is achieved for ANFIS-SVM proposed method using non-reduced features.

Another evaluation is performed on feature space to find the feature sets for ANFIS-SVM classification. The features are classified as spectral feature set that includes "theta ratio" and "alpha ratio," and nonlinear features namely that include "entropy," "standard deviation" and "fractal dimension." **Table 3** shows the results for classification of pain and no-pain condition. It could be observed that non-linear features result into about 17% improvement for SVM and 14% improvement for ANFIS-SVM classification.

6. Discussion

The aim of this chapter was to introduce a classification method base on ANFIS-SVM method for identification of pain condition in EEG signal. In this study, we explored the effectiveness of the identification of pain level and localization of the signals on cortex for therapeutic use [12]. The extracted features of EEG including standard deviation, theta ratio, alpha ratio, entropy, Lyapunov, and fractal dimension and the recording channels in pain EEG signals are studied. The classification method is optimized to identify acute pain. The results of the experiments show that non-linear features combined with the proposed classification method are capable of effective classification. The feature vector is built by entropy, fractal dimension and conventional spectral features. The results also show that the reduction of the number of features could improve the accuracy of the system. Therapeutic usage of this system would be beneficial with patients with anesthesia and the patients who are unable of regular communication.

Author details

Vahid Asadpour[1*], Reza Fazel-Rezai[2], Maryam Vatankhah[3] and Mohammad-Reza Akbarzadeh-Totonchi[4]

1 Kaiser Permanente Southern California, Pasadena, CA

2 American Public University, Charles Town, WV

3 City University of New York, New York

4 Electrical Engineering and Computer Engineering Department, Ferdowsi University of Mashhad, Mashhad, Iran

*Address all correspondence to: v.asadpour@gmail.com

References

[1] Misra G, Wang WE, Archer DB, Roy A, Coombes SA. Automated classification of pain perception using high-density electroencephalography data. Journal of Neurophysiology. 2017; **2**(117):786-795

[2] Tayeb Z, Bose R, Dragomir A, Luke E, Osborn NV, Thakor GC. Decoding of pain perception using EEG signals for a real-time reflex system in prostheses: A case study. Nature Scientific Reports. 2020;**10**:1-11

[3] Chouchou F, Perchet C, Garcia-Larrea L. EEG changes reflecting pain: is alpha suppression better than gamma enhancement? Neurophysiologie Clinique. 2021;**51**(3):209-218

[4] Asraf H, Tahir N, Rizam MSB. A comparative study in kernel-based support vector machine of oil palm leaves nutrient disease. Procedia Engineering. 2012;**41**:1353-1359

[5] Shankar K et al. Optimal feature level fusion based ANFIS classifier for brain MRI image classification. Concurrency and Computation: Practice and Experience. 2020;**32**:1-12

[6] Sebastián M, Navascués M, Otal A, Ruiz C, Idiazábal M, Stasi LLD, et al. Fractal dimension as quantifier of EEG activity in driving simulation. Mathematics. 2021;**9**:1311

[7] Anier A, Lipping T, Mel S, Hovilehto S. Higuchi fractal dimension and spectral entropy as measures of depth of sedation in intensive care unit. In: Annual International Conference of the IEEE Engineering in Medicine and Biology Society. San Francisco: IEEE; 2004. pp. 526-529

[8] Kutepov IE, Dobriyan VV, Zhigalov MV, Stepanov MF, Krysko AV, Yakovleva TV, et al. EEG analysis in patients with schizophrenia based on Lyapunov exponents. Informatics in Medicine Unlocked. 2020;**18**:1-7

[9] Derya E. Lyapunov exponents/probabilistic neural networks for ana lysis of EEG signals. Expert Systems with Applications. 2010;**37**(2):985-992

[10] Hallam A, Morley JG, Green AG. The Lyapunov spectra of quantum thermalization. Nature Communications. 2019;**10**:2708

[11] Murty MN, Raghava R. Kernel-based SVM. In: Support vector machines and perceptrons: Learning, optimization, classification, and application to social networks. Springer International Publishing; 2016. pp. 57-67

[12] Tripanpitak K, He S, Sönmezışık I, Morant T, Huang SY, Yu W. Granger Causality-Based Pain Classification Using EEG Evoked by Electrical Stimulation Targeting Nociceptive Aδ and C fibers. Vol. 9. New York: IEEE Access; 2021. pp. 10089-10106

Chapter 4

Multivariate Real Time Series Data Using Six Unsupervised Machine Learning Algorithms

Ilan Figueirêdo, Lílian Lefol Nani Guarieiro and Erick Giovani Sperandio Nascimento

Abstract

The development of artificial intelligence (AI) algorithms for classification purpose of undesirable events has gained notoriety in the industrial world. Nevertheless, for AI algorithm training is necessary to have labeled data to identify the normal and anomalous operating conditions of the system. However, labeled data is scarce or nonexistent, as it requires a herculean effort to the specialists of labeling them. Thus, this chapter provides a comparison performance of six unsupervised Machine Learning (ML) algorithms to pattern recognition in multivariate time series data. The algorithms can identify patterns to assist in semiautomatic way the data annotating process for, subsequentially, leverage the training of AI supervised models. To verify the performance of the unsupervised ML algorithms to detect interest/anomaly pattern in real time series data, six algorithms were applied in following two identical cases (i) meteorological data from a hurricane season and (ii) monitoring data from dynamic machinery for predictive maintenance purposes. The performance evaluation was investigated with seven threshold indicators: accuracy, precision, recall, specificity, F1-Score, AUC-ROC and AUC-PRC. The results suggest that algorithms with multivariate approach can be successfully applied in the detection of anomalies in multivariate time series data.

Keywords: unsupervised learning, pattern recognition, multivariate time series, machine learning, anomaly detection

1. Introduction

Today, the industry is changing by what experts call the "Fourth Industrial Revolution", also called Industry 4.0. This change is strongly associated with the integration between physical and digital systems through, for example, installing sensors. The integration of these environments allows the collection of a large amount of acquired data in different fields such as: industrial processes, meteorological monitoring stations, stock exchanges etc. This amount of both collected and stored data enables faster and more directed information exchange [1].

In many fields, it is essential for the process to identify unusual patterns that can be generated by unpredictable or unwanted behavior. These behaviors may be due to some problem that may be occurring in the related process, for example, in an industrial environment companies can use machine monitoring data to identify malfunction operating due its abnormal behaviors. This fact, when it is not detected in time, can generate false data and lead experts to misinterpret the operating condition of the machine. Another example would be a credit card operator who can monitor each user's transaction to look for unusual behavior that could point to fraudulent transactions. These unwanted and abnormal behaviors are often called interest patterns and can be extracted from data due a variety of reasons, all presenting a certain level of relevance to the analyst. It is important that this analysis takes into account any changes in the behavior of the parameter to identify opportunities to improve, prevent or correct any situation [2].

The detection of interest/anomaly patterns is usually carried out by specialists which comprises the dynamics of the system under analysis. However, it is often not feasible to analyze and label them due to the large volumes of data generated. Thus, there is a limitation regarding the ability of specialists to process a large amount of data, requiring many hours of work that, in general, are involved in other activities and do not have the time necessary for this relevant activity. Thereby, there is a great need to automate the process of identifying hidden interest patterns in time series data [3].

Unsupervised machine learning (ML) has been research hotspot in intelligence artificial (IA) field to extract useful features from unlabeled raw data. Instead of selecting features by a human operator, the unsupervised learning is quite intelligent and independent of specific knowledge of processing techniques and field expertise in a data-driven way. Thus, there is no escape from the requirement of labeled data to train classifiers at the phase of diagnosis problems, but it is hard to label a mass of collected data before the determination of interest patterns [4].

In an industrial environment, data collection is often carried out through multiple sensors due the possibility of a more robust representation of the phenomena involved, as example, the monitoring of industrial assets is carried out through both acquisition of vibration and temperature data of the machine. Another example is the monitoring of meteorological conditions, which generally collect data on wind speed, air humidity and ambient temperature. However, multivariate data presents a greater challenge for the application of Machine Learning (ML) algorithms, as they must be able to recognize patterns and predict behaviors in a greater amount of data and attributes to be correlated [5].

Anomaly detection algorithms seek for patterns in data that do not conform to an expected behavior. Anomaly detection is essential in industrial applications to optimize economic performance and minimize safety risks. The advent of system health monitoring methods was realized to preserve system functionality within harsh operational environments. Hence, to develop an anomaly detection model based on multi-sensor signals, three major challenges must be faced [6]:

i. online multi-sensor signals are often available in the form of complex, multivariate time series, as different sensors measure various aspects of data over fairly long periods of time, and since there is a large amount of heterogeneous data, it can become impractical to human specialists to label anomalies or unknown events within the data;

ii. especially in industrial applications, it is likely to have extremely imbalanced datasets, since far more data is obtained during normal operations rather than abnormal;

iii. the dataset may contain uncertainties and spurious data, especially when it involves manually recorded data.

One way to mitigate these problems is to perform some type of anomaly detection technique, which are usually computationally intensive algorithms, and then flag unusual patterns for further inspection by human specialists [3]. Other way of dealing with it are based on supervised machine learning anomaly detection algorithms, which require a training dataset that contains a set of instances of anomalies, and a set of instances of non-anomalous (or normal) data, at least. From the training data, the algorithm learns a model that distinguishes between the normal and the anomalous patterns. Such supervised learning algorithms typically require tens or hundreds of thousands of labeled samples to obtain good quality performance. Nevertheless, as stated before, the scarce availability of labeled data poses a challenge to the usage and application of supervised learning techniques for anomaly detection in multivariate time series data.

Hence, unsupervised learning algorithms step up as a viable and feasible alternative to tackle this challenging problem. Since they are designed to deal with unlabeled data, they are able to learn and identify interesting patterns from the data's own internal structure, meaning that they can be used to point out anomalous patterns when the labels are unknown. Thus, unsupervised machine learning models are essential to solve the addressed challenges [7].

Several works proposed the development and application of unsupervised ML algorithms over the past years to detect anomalous patterns in time series datasets, which are based on major approaches that are summarized in **Table 1**. A detailed description and evaluation of each of these approaches is beyond the scope of this chapter.

Most of the works based on unsupervised ML algorithms employs clustering techniques, which are either distance-based [10] or density-based [11]. On the one hand, the advantages of clustering methods are that they are simple, robust, and easy to program. However, the problem is the need to define the parameters related to the data observations beforehand such as defining a similarity function or the number of clusters that should exist in the data, and that becomes the responsibility of the designer to determine how these parameters should be used, even if the data has a random structure [9]. The work [12] proposed the K-means to automate diagnosis of defective rolling bearing. To overcome the sensitivity of choosing the initial clusters number, the initial centers were selected using features extracted from simulated signals. However, K-means depends mainly on distance calculation between all data points and the centers, therefore, the cost of the computational time will be higher for big data.

To reduce the time cost of K-means, [13] proposed a Fast K-means algorithm based on two stage. The first stage is a fast distance calculation using only a small fraction of the data to originate the best possible location of the centers. The second stage is a slow distance calculation in which the initial centers are taken from the first stage. Besides that, the K-means is optimized through grid search method that is efficient when the number of parameters is small.

Aiming to prove the reduction quality of the dataset, [14] demonstrated the superiority of the autoencoder in the feature dimensional reduction comparing with the PCA method. The reduced feature set obtained from the PCA method revealed overlaps of classes and features that are scattered on the large space, while the autoencoder represented the superior ability in the clear and concentrated distribution.

The work [15] presented a similar comparative study to evaluate the performance of the autoencoder to the original feature set, PCA reduction and real-valued negative selection. The dimensional reduction of the autoencoder, once again, have performed a highly improved anomaly detection compared to the others. Moreover, PCA space transformation requires complete knowledge about normal and faulty data classes.

Algorithm	Description
K-means	It is used to divide a group of data points into hard clusters. It assumes a balanced cluster size, the joint distribution of features with equal variance, and independent features with similar cluster density. Determining the optimal K can be difficult, but for small values, it is computationally fast and efficient. In addition, it is important to choose the most appropriated distance or similarity function, since it is one of the key aspects used to determine whose cluster a certain data point belongs to.
Gaussian mixture model	It uses a Gaussian distribution-based parametric model to identify the underlying populations. These can be explained by a normal distribution in the midst of many heterogeneous populations. However, in many practical situations, the data distribution may not have any explicit clusters. As a result, each point can be assigned with different weights or probabilities to soft clusters.
Random forest	It operates by constructing a multitude of decision trees at training time and outputting the class that is either the mode of the classes (classification) or the mean prediction (regression) of the individual trees. Random forest can learn arbitrary relationships between the features and the outcome, even non-monotonic relationships.
k-NN	It assigns data points according to the majority of its nearest neighbors to find anomalous data points by measuring the local deviation. A choice needs to be made on the value of K, i.e. the number of neighbors, to avoid overfitting/underfitting issues.
DBSCAN	It recognizes the clusters as dense regions having some coincidence that is diverse from the other sparse region. The algorithm may use a reduced number of points and a distance measure to merge the data points that are similar to each other. Moreover, DBSCAN requires two parameters to operate, which are the epsilon (*eps*) and the minimum points (*minpoints*). *Eps* determines the smallest distance existing between two points in a cluster, while *minpoints* defines the least number of points required to form a dense region [8].
PCA	Principal Component Analysis (PCA) based anomaly detection techniques are able extract the main features of a certain dataset without losing its ability to represent the original data, then using these features to analyze which constitute a normal class and applies distance metrics to identify cases that represent anomalies. This allows to train a model using existing imbalanced data.
Autoencoders	It is a neural network that attempts to reconstruct its input. Similarly to PCA, it can serve as a form of extract main features to produce a compressed representation of its input at the encoder. This representation can be mapped to its original form using a decoder. Anomaly detection uses the reconstruction error to measure how well the decoder is performing.

Adapted [9].

Table 1.
Unsupervised ML approaches found in literature for anomaly detection in time-series.

Two methods for feature selection were proposed by [15]. The first one is based on k-NN for clustering using feature similarity influence, and the second one is the pretraining using sparse autoencoders. The classification performance obtained by the k-NN algorithm is comparable to the result obtained from the autoencoder (slightly lower in accuracy). The criterion for choosing the "k" parameter is based on the combinations that frequently appear in the subsets reduced by the technique itself. In other words, the criterion adopted is purely empirical.

Furthermore, even though several unsupervised techniques have been proposed in literature, their performance depends a lot on the data and application they are being used in. This indicates that most of these methods have little systematic advantages over the other when compared across many other datasets.

In this context, this chapter discuss the level of accuracy and reliability of six unsupervised ML algorithms for pattern recognition and anomaly detection with no need of labeled data. Two real cases were applied for performance evaluation of the algorithms abilities to detect the interest patterns in the multivariate time series data.

The real cases were: (i) meteorological data from a hurricane season and (ii) monitoring data from dynamic machinery for predictive maintenance purposes.

2. Unsupervised ML algorithms

This section will review the concept and application of six unsupervised ML algorithms for anomaly/pattern detection applied in this research. It is important to inform that among the six algorithms described in this chapter, only the methods in Sections 2.1 and 2.3 has its intrinsic characteristic to perform a multivariate analysis, while the other algorithms are only able to perform a univariate analysis. Therefore, a univariate performance evaluation is computed in each dimension of the data to then calculate an average performance, except algorithms in 2.1 and 2.3 sections.

2.1 C-AMDATS

The Cluster-based Algorithm for Anomaly Detection in Time Series Using Mahalanobis Distance (C-AMDATS) is a clustering ML unsupervised algorithm. The model has only two hyperparameters that user can manipulate: (i) Initial Cluster Size (ICS) and Clustering Factor (CF). First the ICS clusters the observed sequences of time series data A, where each cluster may represent a behavior status. After the initial clustering, a new and better clustering in the dataset is remade according to the data points distribution over timeline. This ability is due to the usage of the Mahalanobis distance in the algorithm. In general, clustering techniques use the Euclidean distance function, which makes the clustering assumes the geometric shape of a circle, then it does not consider the variance of each dimension or feature of the dataset. However, there are situations in which the variance between each dimension (or feature) is different. Conversely, by using the covariance matrix, the Mahalanobis distance can detect the variance of each dimension. Eq. (1) presents the Mahalanobis distance formula.

$$d_m(x,\mu)=\sqrt{(x-\mu)^T S^{-1}(x-\mu)} \tag{1}$$

Where: $d_m(x,\mu)$ is the Mahalanobis distance between a specific point in the time series and its respective centroid; $x = (x_1, x_2, ..., x_n)^T$ is a specific variable in the time series data, where n is the number of variables; $\mu = (\mu_1, \mu_2, ..., \mu_n)^T$ is a certain cluster centroid; and S is the covariance matrix relative to that cluster.

After the new clustering through the Mahalanobis distance, the algorithm calculates the similarity of each cluster in the time series A to find the respective hidden patterns P. This similarity is calculated using the standard deviation σ_y of the actual values of the A samples, the Y coordinate of each centroid and the CF. If the modulus of the difference between the y coordinate of the centroids of two cluster is less than or equal to the product of CF and σ_y, then these clusters can be merged, meaning that they will represent the same pattern P. This task is carried out until every cluster have been analyzed.

The last step of C-AMDATS is to calculate the probability of the pattern P to be an anomaly R. The Anomaly Score measures the anomaly R for each pattern P (found in the previous step). The score is calculated by the ratio of the size of the entire time series to the sum of the sizes of the clusters present in P. The anomaly score assesses the degree of relevance of P in terms of anomaly detection. Then, all set P is ordered by R in descending order, and the anomalous patterns will be those with the highest anomaly score values. The higher the anomaly score value for a pattern P, the greater probability is of being an anomaly behavior in A [2]. Eq. (2) presents the Anomaly Score formula.

$$Anomaly\ Score_{P_i} = \frac{|T|}{|P_i|} \tag{2}$$

Where $Anomaly\ Score_{P_i}$ is the anomaly score of the pattern P_i, $|P_i|$ is the size of the pattern P_i, and $|T|$ is the size of the time series T.

2.2 Luminol Bitmap

Bitmap is an available unsupervised learning algorithm in Luminol library for anomaly detection or time series correlation. The background of Bitmap algorithm is based on the idea of time series bitmaps. The logic of the algorithm is to make a feature extraction of the raw time series data - by converting them into a Symbolic Aggregate Approximation (SAX) representation - and use it to compute the information about the relative frequency of its features to color a bitmap in a principled way. SAX allows a dimensionality reduction of the raw time series C of arbitrary length n to a string arbitrary length w ($w < n$, typically $w << n$) by a vector $\bar{C}$. It transforms the data into a Piecewise Aggregate Approximations (PAA) representation to symbolize it into a discrete string [16]. Eq. (3) presents the calculation of the *ith* element of $\bar{C}$:

$$\bar{C}_i = \frac{w}{n} \sum_{j=\frac{n}{w}(i-1)+1}^{\frac{n}{w}i} C_j \tag{3}$$

After transformed a time series dataset into PAA, the algorithm applies a further transformation to obtain a discrete representation with equiprobability [16]. The conversion of the time series into a SAX words is made by a slider window (also called feature window). Bitmap algorithm use two concatenated slider windows together across the sequence, the latter one is called lead window, showing how far to look ahead for anomalous patterns and the former one is called lag window, whose size represents how much memory of the past to remember it.

In summary, the algorithm approach is to convert both feature windows into SAX representation, then count the frequencies of SAX subwords at the desired level and get the corresponding bitmaps. The distance between the two bitmaps is measured and reported as an anomaly score at each time instance, and the bitmaps are drawn to visualize the similarities and differences between the two windows. The user must choose the length of the feature windows N and the number n of equal sized sections in which to divide N [3].

2.3 SAX-REPEAT

SAX-REPEAT algorithm is an approach that relies on extending the original SAX implementation to handle multivariable data. The algorithm takes as input a set of K multivariable time series X_i of lengths T_i, and dimensionality D, that represent different instances of the raw data to be learned. The user can set the parameters of the final string length N and an alphabet size M.

The algorithm applies SAX to each dimension of the data separately, and then combine the output string by assigning each possible combination of symbols, resulting in D strings to a unique identifier. This leads to a string of length N, but an extended alphabet of length M^D. So, to maintain the requirement of the final string to be an alphabet of symbols M (parameter set by the user), the algorithm clusters the resulting characters into M clusters through K-means method and replace each character with the centroid of its cluster [17].

Although SAX-REPEAT can recognize interesting patterns, the original algorithm does not calculate the probability of the patterns being anomalous. Thereby, this work implemented an anomaly score for each found cluster (pattern). As C-AMDATS algorithm, the score is computed by the ratio between the size of the entire time series and the sum of the sizes of each cluster. Therefore, each cluster is sorted according to the respective anomaly score in descending order, and the anomalous patterns will be those with the highest anomaly score values.

2.4 k-NN

The k-Nearest Neighbors (k-NN) algorithm is one of the most popular method to solve both classification and regression problems. However, in this study, we will use it only for classification problem as unsupervised learning.

The algorithm assumes that similar data points exist in proximity, *i.e*, they are near to each other. The algorithms capture the idea of the similarity (also known as distance, proximity, or closeness), calculating the distance between points on a graph. Distance calculation is usually done by Euclidean distance, but it can be calculated using other distance functions. The Euclidean distance between the points $P = (p_1, p_2,...,p_n)$ e $Q = (q_1, q_2,...,q_n)$, in a n-dimensional Euclidean space, it is defined in Eq. (4):

$$d_e(p,q) = \sqrt{\sum_{i=1}^{n} (p_i - q_i)^2} \tag{4}$$

Where, $p = (p_1, p_2, ..., p_n)$ and $q = (q_1, q_2, ..., q_n)$ are two points in Euclidean n-space.

The k-NN algorithm depends on two parameters, a metric used to compute the distance between two points (in this case Euclidean function), and a value k of the number of neighbors to consider. When k is underestimated, the algorithm can overfit, *i.e.* it will classify just based on the closest neighbors instead of learning separating frontier between classes, but if k is overestimated, the algorithm will underfit, in the limit if $k = n$, the algorithm will consider every point belongs to the class that has more samples [18, 19].

2.5 Bootstrap

Bootstrap algorithm uses the computational power to estimate almost any summary statistics, such as the confidence interval, mean, or standard error. The method depends on the notion of a bootstrap sample B, which is a resampling of size n drawn to replace the original dataset $Z = (Z_1, Z_2, ..., Z_n)$. The bootstrap sample is represented $Z^* = (Z_1^*, Z_2^*, ..., Z_n^*)$. Each Z_i^* is one of the original Z values randomly selected, the selection probability for each Z value is equipollent, for example: $Z^* = Z_7, Z_2^* = Z_5, Z_3^* = Z_9, Z_4^* = Z_7$, etc. Note that the same original value can appear zero, one or more times, in the example, Z_7 appeared twice, *i.e*, the selection of Z value is not exclusive. The name Bootstrap concern to the use of the original dataset to generate new datasets Z^*. The idea is to generate a larger number of Bootstrap sample B of each size n using a random number device to perform the algorithm training. The number of bootstrap repeats defines the variance of the estimate, *i.e*, higher the number is, better is the variance, but in contrast, the computational cost increases with the increasement of the B number [20, 21].

In this sense, we are interested in calculating a confidence interval using Bootstrap, which is performed by requesting the statistics stored during the training and selecting values in the chosen percentile for the confidence interval. The chosen percentile is denoted as δ (Alpha or Significance Level). Eq. (5) defined the calculation to estimate the distribution of δ^* for each Bootstrap sample.

$$\delta^* = \overline{x}^* - \overline{x} \tag{5}$$

Where: $\overline{x}^*$ is the mean of an empirical bootstrap sample and $\overline{x}$ is the mean of the original data.

Therefore, the confidence interval for a Significance Level of 0.05 is defined by Eq. (6).

$$\text{Confidence interval} = \left[\overline{x} - \delta^*_{.05}, \overline{x} - \delta^*_{.95}\right] \tag{6}$$

Where, $\overline{x}$ is the mean of the original data, $\delta^*_{.05}$ is significance level at the 5th percentile, and $\delta^*_{.95}$ is significance level at the 95th percentile.

So, in order to obtain a very accurate estimate of $\delta^*_{.05}$ and $\delta^*_{.95}$, it is important to generate a large number of bootstrap samples.

2.6 RRCF

Robust Random Cut Forest (RRCF) algorithm is an ensemble technique for detecting outliers. The idea is based on an isolation forest algorithm that uses an ensemble of trees. In graph theory, trees are collections of vertices and edges where any two vertices are only connected by one edge, it is an ordered way of storing numerical data.

In this view, the algorithm takes a set of random data points, cuts them to the same number of points and creates trees. The algorithm starts by constructing a tree of n vertices, then it creates more trees of the same size, which in turn creates the forest. The user can choose the number of trees and the number of data point of each tree has, which is randomly sampled from the dataset. After the construction of the forest, the algorithm injects a new data point p into the trees to follow the cuts and to compute the average depth of the point across a collection of trees. The point is labeled an anomaly if the score overtake a threshold, which corresponds to the average depth across the trees [22].

3. Comparative analysis between the algorithms

This section will discuss the details of the datasets, the algorithm parameters settings, the evaluations performance method, and a comparative analysis between the algorithms.

In order to summarize the advantages and limitations of each algorithm, **Table 2** shows advantage and limitations of the ML algorithms.

The advantages and limitations of each algorithm will be discussed throughout the development of this chapter.

3.1 Characterization of case studies

This chapter brings two sets of real data, which were collected for the purpose of detecting anomalies in multivariate time series. The databases applied in this study will be detailed in the next sections.

ML algorithms	Advantage	Limitations
C-AMDATS	• Multivariate approach • Easy to parameterize • Considers the variances of each dimension	• Take more CPU time due to the need to compute the inverse of the covariance matrix for each cluster • Sensitive to noisy data, missing values and outliers
Luminol Bitmap	• Quick calculation time • Runs well on big data • Not too sensitive to parameter choices	• Univariate approach
SAX-REPEAT	• SAX Multivariate approach	• Hard to parameterize • Slow calculation time due the k-means clustering • Sensitive to missing values and outliers
k-NN	• Quick calculation time • Easy to implement • Variety of distance criteria can be chosen	• Does not work well with large dataset • Does not work well with high dimensions • Sensitive to noisy data, missing values and outliers • Univariate approach
Bootstrap	• Quick calculation time • Does not require large sample size	• Can be computationally expensive depending on the bootstrap sample number • Univariate approach
RRCF	• Different dimensions are treated independently • Designed to run in a streaming data	• Univariate approach • Can be computationally expensive depending on the number of trees • Sensitive to noisy data, missing values

Table 2.
Advantage and limitations of the unsupervised ML algorithms.

3.1.1 Case study 01 - meteocean data in hurricane season

The chosen set is a public meteocean data available online in the National Data Buoy Center of the National Oceanic and Atmospheric Administration's (NOAA). The dataset was collected in the Atlantic Ocean off the Bahamas coast (23,838 N; 68,333 W). The data were structured in hourly frequency and it begins in June 2012 until November 2012 (213 days and 22 hours), comprising 15,315 data points. This period corresponds to the hurricane season in the Ocean Atlantic, which that year was especially active with 19 tropical cyclones (winds above 52 km/h), which 10 cyclones became hurricanes (winds above 64 km/h).

Hurricanes can be detected by several meteorological variables that consequently are directly impacted. In this case study, the following analysis variables were considered: (a) significant wave height (WVHT); (b) sea level pressure (PRES); and (c) wind speed (WSPD). Within the period of the dataset, three hurricanes transited through the Bahamas coast region: (i) Isaac; (ii) Rafael and (iii) Sandy.

Isaac had his first alert issued on August 21 by the National Hurricane Center. Several islands in the Lesser Antilles have been placed under hurricane surveillance or tropical storm warnings. Isaac was tracked between Guadalupe and Dominica on August 22, it passed over Haiti and Cuba with a strong tropical storm force.

On August 26, the Isaac approaches Florida Keys and the next day entered the eastern Gulf of Mexico causing several economic impacts in the USA. There was a gradual intensification and Isaac reached its peak intensity as a category 1 hurricane, with sustained 1-minute winds of 80 mph (130 km/h) [23].

Hurricane Rafael produced minor damage in the northeastern Caribbean Sea in mid-October 2012. The first alert was issued to Bermuda on October 14, but was canceled on October 17 when the hurricane passed northeast of the island. On October 16, Rafael reached his peak intensity with maximum sustained winds of 90 mph (150 km/h). Rafael intensified in a category 1 hurricane [24].

Hurricane Sandy was the deadliest and most destructive, as well as the strongest, hurricane of the 2012 Atlantic hurricane season. Inflicting nearly $70 billion USD in damage, Sandy was a Category 3 storm at its peak intensity when it made landfall in Cuba. On October 24, Sandy became a hurricane reaching the coast near Kingston and Jamaica. On October 25, it hit his peak intensity in Cuba. On October 31, Sandy was already off the coast of Maine in the United States of America [25].

The **Figure 1** illustrates the period of hurricanes Isaac, Rafael and Sandy in the multivariate time series data.

In **Figure 1**, it is possible to visualize a behavior similarity between the three hurricanes. During the passage of the hurricanes, the variables WVHT and WSPD presented upward spikes, but on the other hands, PRESS presented downward spikes.

It is worth noting that the period of hurricanes in the **Figure 1** represents the time of its trajectory on the coast of Bahamas, and not its life span throughout its trajectory in the Atlantic Ocean.

3.1.2 Case study 02: monitoring data from dynamic machinery

The public dataset provided by the KNIME was acquired from 28 sensors installed in a dynamic machine. The sensors were installed to collect eight mechanical components parts (1st column of **Table 3**). The data starts on January 1st of 2007 and goes until April 20th of 2009 (838 days), comprising 16,660 data points.

The dataset was composed of 28 time series from 28 sensors. The signals were pre-processed with Fast Fourier Transform (FFT). The **Table 3** shows the groups and description of the sensors.

Each sensor group had at least 3 collections with different frequency bands, except the torque variable (M1), which had only one collection.

Signs of rotor malfunction could be traced back to March 6, 2008. The breakdown event happened on July 21, 2008. The break was visible only to some sensors, especially with low frequency bands.

For a cleaner and clearer view, **Figure 2** illustrates the multivariate time series only for sensors that detected the malfunction zone of the dynamic machine. Therefore, of total of 28 sensors, 18 were chosen to illustrate the multivariate time series. The machinery malfunction was detected in all sensor groups, except for the M1.

In **Figure 2**, it is possible to verify a behavior change in the multiple sensors inside the malfunction zone (begging of March until the end of July). The **Figure 2** also illustrates two alarms in the beginning of 2007 triggered by the KNIME system. However, these two alarms can be considered as pre-mature, as the history of the machinery continued to run normally over one year. Another detail is that, afterwards the breakdown and rotor replacement, the signals were recorded much cleaner.

3.2 Parameterization of the ML algorithms

The parameters of unsupervised ML algorithms were settings to achieve the best possible performance to find the patterns of interest. The parameterization requires

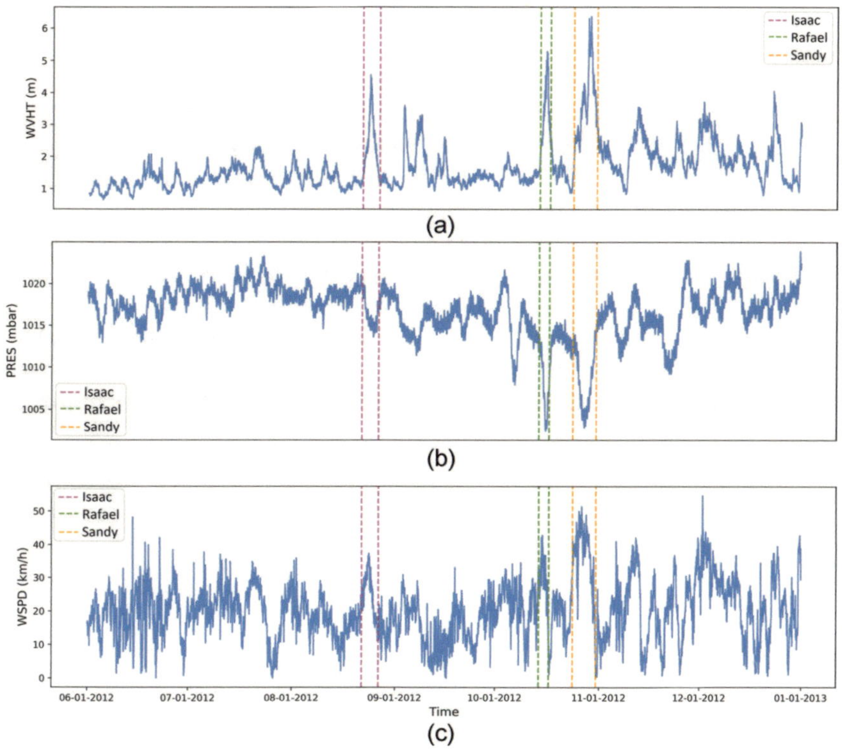

Figure 1.
Visualization of three variables in the same time domain (a) significant wave height, (b) sea level pressure and (c) wind speed. Color boxes represent hurricanes Isaac, Rafael and Sandy. Source: produced by the authors.

Sensor Group	Sensor Description
A1	Input shaft vertical
A2	Second shaft horizontal upper bearing
A3	Third shaft horizontal lower bearing
A4	Internal gear 275 degrees
A5	Internal gear 190.5 degree
A6	Input shaft bearing 150
A7	Input shaft bearing 151
M1	Torque kNm

Table 3.
The eight parts of the rotor monitored through groups of sensors.

several attempts of success and error to achieve the best possible result. SAX-REPEAT was the most difficult method of setting the parameters due the high sensitivity of the variables. Whereas the Luminol Bitmap revealed not too sensitive to parameter choices, where the Bitmap Detector Score demonstrated the most determinant parameter for the algorithm. The Bootstrap showed a similar result for iterations above 200 and confidence level above 95%. C-AMDATS, RRCF and k-NN are easy algorithms to set the parameter due the small number they have. As an example of experiment case 2, it was necessary to run SAX-REPEAT with 76 different combinations of parameters to identify the best configuration, k-NN was necessary to run 21 times, C-AMDATS 20 times, RRCF 11 times, Luminol 12 times, and Bootstrap 10 times.

The **Table 4** summarizes the parameter settings of the presented algorithms for the two real cases applied in this chapter.

All ML algorithms in this paper were implemented in Python 3.6 programming language and executed on a high performance computing named AIRIS (Artificial Intelligence RSB Integrates System) at the Supercomputing Center for Industrial Innovation at SENAI CIMATEC. The AIRIS processor model is the Intel(R) Xeon(R) Gold 6148 CPU @ 2.40GHz and has 376 GB RAM memory.

3.3 Case study experiment 01 - meteocean data in hurricane season

All monitoring variables at the meteocean data were processed in ML algorithms using the settings presented in the **Table 4**. The results were compared to the period of hurricanes life as shows in **Figure 1**. The hurricanes behaviors are

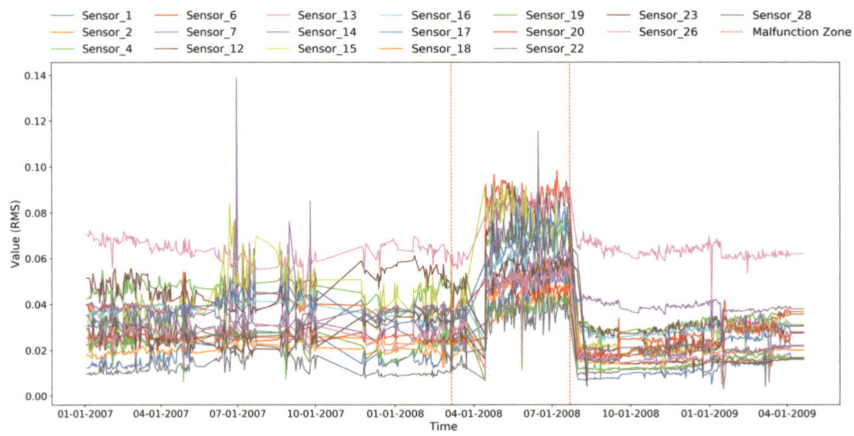

Figure 2.
Multivariate time series of 18 sensors that detected the malfunction zone of the machine.

Algorithm	Parameter Setting	
	Case 01	**Case 02**
C-AMDATS	CF = 3.0 ICS = 24	CF = 1.8 ICS = 30
Luminol Bitmap	Bitmap Detector Score = 1.0 Precision = 40 Lag Window Size = 10 Future Window Size = 10 Chuck Size = 24	Bitmap Detector Score = 0.35 Precision = 40 Lag Window Size = 10 Future Window Size = 10 Chuck Size = 30
SAX-REPEAT	Window size = 1155 PAA size = 3 Alphabet length = 2	Window size = 120 PAA size = 3 Alphabet length = 3
k-NN	k = 5 Metric = Euclidean Distance	k = 150 Metric = Euclidean Distance
Bootstrap	Confidence = 0.95 Number of Iterations = 200	Confidence = 0.95 Number of Iterations = 200
RRCF	Number of Trees = 10 Tree Size = 5105	Number of Trees = 20 Tree Size = 595

Table 4.
Parameter setting of the Unsupervised ML algorithms. CF: Cluster Factor - ICS: Initial Cluster Size - PAA: Piecewise Aggregate Approximations - k: Neighbors number.

more visually clear through the WVHT variable. Thereby, for the better understanding of the reader, we only illustrated the ML results in the WVHT variable, even though was made a multivariate analyzed. **Figure 3** shows the detection of the six algorithms.

In **Figure 3**, the C-AMDATS algorithm detected three distinct behavior patterns in the multivariate time series. Patterns 0 and 1 had the highest anomaly score and are well situated in hurricanes regions, so these patterns where considered as anomaly behavior. However, pattern 1 also appear in November and December, which had no records of hurricane or tropical depression or tropical storm in the Bahamas cost, revealing to be a false positive signal. The Luminol bitmap and RRCF algorithms failed to isolate the patterns of interest. Luminol demonstrated a little sensitivity for detecting anomalies in this experiment, because it was possible to verify a few data points detected as anomaly. SAX-REPEAT returned 6 distinct patterns, which patterns 4 and 5 were the top 2 of the anomaly score. These two patterns are precisely in the regions of interest, however, it is possible to verify these patterns also in other regions, also indicating false positive signals. Bootstrap and k-NN had similar results, both algorithms detected spikes caused by hurricanes, but with many false positives, especially Bootstrap.

Therefore, it is possible to ascertain that the algorithms with the best performance in detecting the patterns of interest in case 01 were C-AMDATS and SAX-REPEAT. But a quantitative analysis will still be performed.

3.4 Case study experiment 02 - monitoring data from dynamic machinery

Analogous to the experiment performed in case 1, the experiment case 2 brings the results of the patterns and anomalies detection of unsupervised learning algorithms in the KNIME dataset.

All 28-monitoring data were processed using the settings in **Table 4**. The results were compared to the period of malfunction of the machine as show in **Figure 2**.

The machine malfunction is more visually clear through the sensor 1. Therefore, **Figure 4** only illustrated the results of Sensor 1, although the analysis was performed in a multivariable way.

In **Figure 4**, the C-AMDATS algorithm detected three distinct behavior patterns in the multivariate time series. Patterns 0 had the highest anomaly score and is well situated in the interest region, so this pattern was assumed to be anomalous. Luminol and RRCf again failed to isolate the fault, both algorithms had many false positives and false negatives. SAX-REPEAT detected 15 different patterns, which is not desired as it makes difficult for the specialist to analyze many patterns. Nevertheless, patterns 0 and 4 had the lowest punctuation in the anomaly score ranking, so these patterns were assumed to be normal and the others as anomaly. The k-NN and Bootstrap methods also demonstrated a good performance in isolating the period of interest, with few false positives and false negatives.

Therefore, the algorithms were able to isolate the anomalous region well in case 02, with exception of Luminol Bitmap and RRCF.

3.5 Performance evaluation

The performance evaluation of the algorithms - in their ability to identify the same anomalous patterns - was performed through the calculation of seven metrics: accuracy (ACC), precision (PR), recall (REC), specificity (SP), F1-score (F1), area under the curve (AUC) of receiver operating characteristics (AUC-ROC), and AUC of precision and recall curve (AUC-PRC).

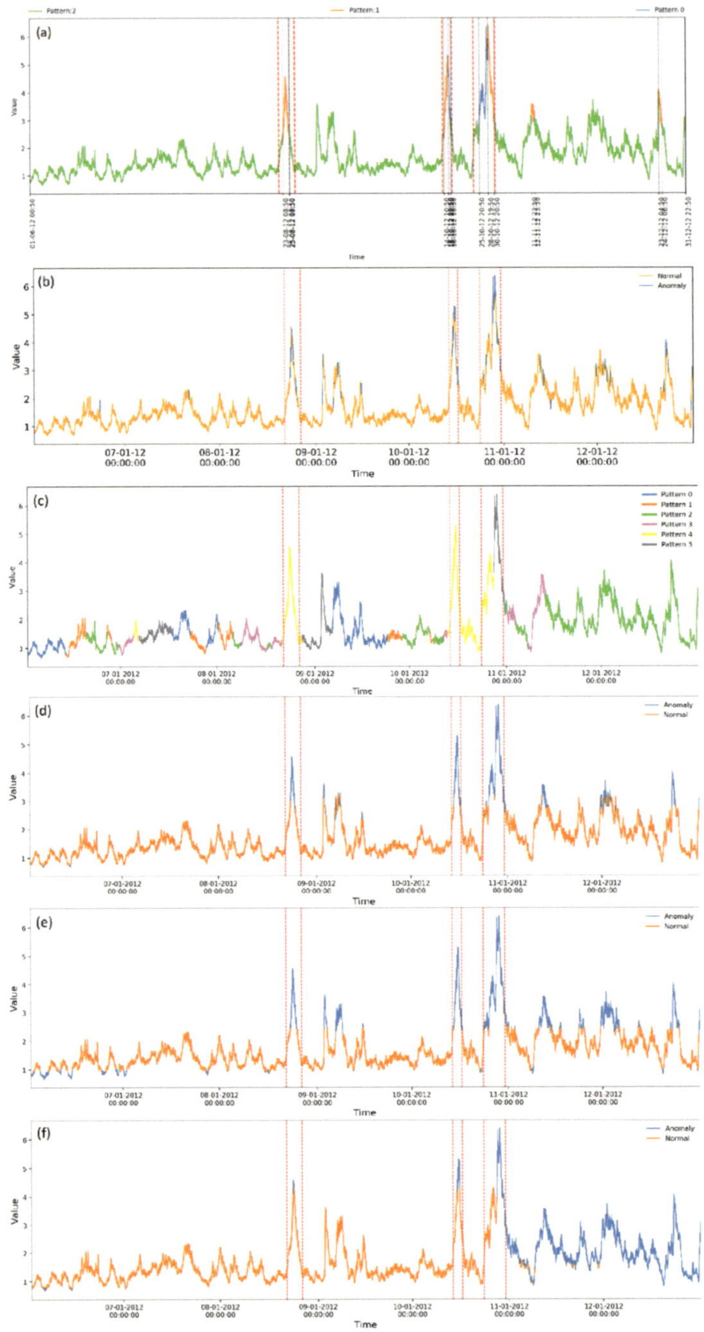

Figure 3.
Results of the unsupervised algorithms of Case 1. (a): C-AMDATS, (b): Luminol Bitmap, (c): SAX-REPEAT, (d): K-NN, (e): Bootstrap, and (f): RRCF.

However, the performance evaluation would not be properly fair, as the Luminol, k-NN, Bootstrap and RCCF algorithms made the analysis univariably (different from C-AMDATS and SAX-REPEAT). Thereby, in an attempt to obtain a more appropriate analysis, the threshold metrics was calculated for all proposed variables and then extracted an average evaluation, except C-AMDATS and SAX-REPEAT.

All the evaluation metrics are calculated by comparing the real data points (classified by experts) with the predicted data points (predicted by ML algorithms). So, the ACC reveals the correct prediction in a general approach, but it may hide the error rate of the model, that is why it is prudent to measure the performance jointly with other metrics. PR indicates the true positive value compared to the false negative. REC reveals out the true positive value with the false positive.

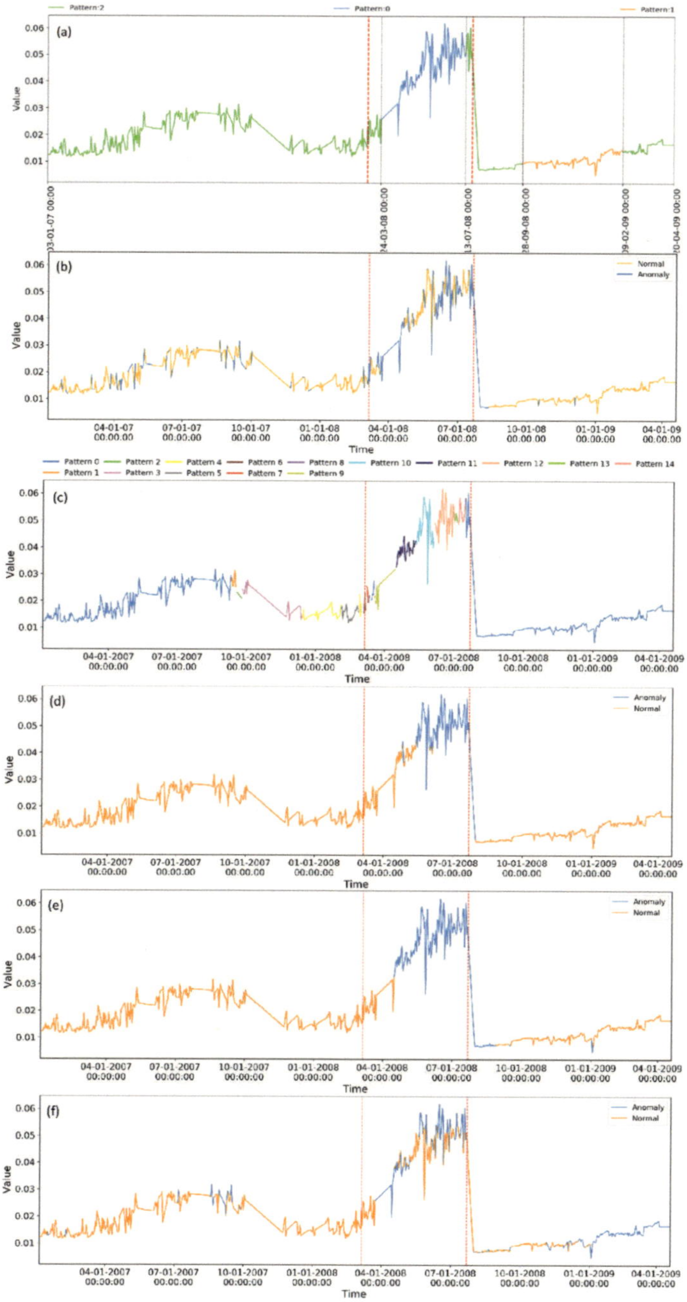

Figure 4.
Results of the unsupervised algorithms of Case 2. (A): C-AMDATS, (B): Luminol Bitmap, (C): SAX-REPEAT, (D): K-NN, (E) Bootstrap, and (F): RRCF.

ML Algorithms	Metrics	Case #1	Case #2	Average
C-AMDATS	ACC	96%	96%	96%
	PR	90%	98%	94%
	REC	80%	89%	85%
	SP	80%	89%	85%
	F1	84%	92%	88%
	AUC-ROC	81%	89%	85%
	AUC-PRC	76%	88%	82%
Luminol Bitmap	ACC	86%	72%	79%
	PR	55%	58%	57%
	REC	55%	61%	58%
	SP	55%	61%	58%
	F1	55%	58%	57%
	AUC-ROC	56%	61%	58%
	AUC-PRC	52%	56%	54%
SAX-REPEAT	ACC	86%	89%	88%
	PR	66%	82%	74%
	REC	92%	90%	91%
	SP	92%	90%	91%
	F1	70%	85%	78%
	AUC-ROC	92%	90%	91%
	AUC-PRC	66%	79%	73%
k-NN	ACC	94%	86%	90%
	PR	80%	81%	80%
	REC	67%	68%	67%
	SP	67%	68%	67%
	F1	69%	71%	70%
	AUC-ROC	66%	68%	67%
	AUC-PRC	61%	67%	64%
Bootstrap	ACC	79%	85%	82%
	PR	59%	72%	65%
	REC	73%	75%	74%
	SP	73%	75%	74%
	F1	59%	73%	66%
	AUC-ROC	73%	75%	74%
	AUC-PRC	56%	70%	63%
RRCF	ACC	80%	66%	73%
	PR	55%	44%	50%
	REC	55%	45%	50%
	SP	55%	45%	50%
	F1	54%	44%	49%
	AUC-ROC	55%	45%	50%
	AUC-PRC	53%	49%	51%

Table 5.
Performance Evaluation of unsupervised ML algorithms to detect interesting/anomalous patterns in multivariate time series data.

Both metrics (PR and REC) reveal the model's ability to predict positive values, but with different perspectives. SP demonstrated the capacity of the model to predict the true negative over false positives perspective. The F1 is a harmonic average between REC and PR. AUC-ROC is the area under the curve on the true positive (REC) and false positive (1- SP) rates. The AUC-PRC is the area below the curve between PR and REC. AUC-PRC is an important metric for assessing unbalanced datasets, being a great advantage over the others, since in the vast majority of cases, especially real data, have a higher volume of normal than abnormal data.

The seven performance assessment metrics for all proposed variables of case 1 and case 2 experiments are listed in the **Table 5**.

The performance evaluation presented in the **Table 5** revealed that the C-AMDATS was the one that stood out amongst the other algorithms. C-AMDATS was superior in ACC, PR, AUC-PRC and F1 metrics against SAX-REPEAT, which was the second algorithm that stood out. Nevertheless, it is relevant to note that C-AMDATS was 10% superiority in AUC-PRC of SAX-REPEAT. Then, in decreasing order of algorithm position in the performance evaluation would be: (i) C-AMDATS, (ii) SAX-REPEAT, (iii) k-NN, (iv) Bootstrap, (v) Luminol and (vi) RRCF. Both algorithms that have a multivariate analysis intrinsically were superior. However, more case studies must be carried out to affirm the superiority of the algorithms studied here.

Therefore, the results presented in this study strengthens the idea that unsupervised machine learning algorithms can assist the data annotation and labeling process. This approach can optimize much of the specialists' time and leverage the supervised AI models.

4. Conclusions and future work recommendations

This work demonstrated the effectiveness of a multivariate analysis using six different unsupervised ML algorithms for time series. To verify the performance of the unsupervised ML algorithms to detect interesting/anomalous patterns in real time series data, the six algorithms were applied in two different real cases: (i) meteocean data in hurricane season and (ii) monitoring data from dynamic industrial machinery. The experimental results showed that clustering methods as C-AMDATS have higher capacity to recognize and isolate the anomaly region, revealing the ability to assist experts to label raw data with unsupervised ML algorithms with great performance.

Future works include the extension of this analysis to more real cases aiming to develop a broader analysis, as well as an extensive study and investigation of approaches of semi-supervised learning to train deep learning algorithms to predict and classify unknown data in different dataset.

Acknowledgements

The authors thank the NOAA and KNIME for providing to the scientific community publics datasets, leveraging researches and technology development. The authors also thank the research group CS2i and the Reference Center for Artificial Intelligence at SENAI CIMATEC.

Author details

Ilan Figueirêdo, Lílian Lefol Nani Guarieiro
and Erick Giovani Sperandio Nascimento*
SENAI CIMATEC, Salvador, Brazil

*Address all correspondence to: erick.sperandio@fieb.org.br

References

[1] Rodner E, Barz B, Guanche Y, Flach M, Mahecha M, Bodesheim P, et al. Maximally Divergent Intervals for Anomaly Detection. In: ICML Workshop on Anomaly Detection [Internet]. New York, NY, USA; 2016. Available from: http://arxiv.org/abs/1610.06761%0A

[2] Sperandio Nascimento EG, Tavares O, De Souza A. A Cluster-based Algorithm for Anomaly Detection in Time Series Using Mahalanobis Distance. In: ICAI'2015 - International Conference on Artificial Intelligence [Internet]. Las Vegas, Nevada, USA: ICAI'15 - The 17th International Conference on Artificial Intelligence; 2015. p. 622-8. Available from: https://www.researchgate.net/publication/282330724_A_Cluster-based_Algorithm_for_Anomaly_Detection_in_Time_Series_Using_Mahalanobis_Distance

[3] Wei L, Kumar N, Lolla VN, Keogh EJ, Lonardi S, Ratanamahatana C (Ann). Assumption-Free Anomaly Detection in Time Series. In: 17th International Conference on Scientific and Statistical Database Management, SSDBM [Internet]. Berkeley, CA, USA; 2005. p. 237-42. Available from: https://pdfs.semanticscholar.org/909b/8226968d41f76cc14d0eef2d365572e7a37b.pdf?_ga=2.68813208.804195361.1594613685-826585445.1591805583

[4] Liu H, Zhou J, Xu Y, Zheng Y, Peng X, Jiang W. Unsupervised fault diagnosis of rolling bearings using a deep neural network based on generative adversarial networks. Neurocomputing [Internet]. 2018 Nov;315:412-24. Available from: https://linkinghub.elsevier.com/retrieve/pii/S0925231218308695

[5] Figueiredo IS, Guarieiro LLN, Santos AAB, Nascimento EGS. Algoritmo de aprendizagem de máquina não supervisionado para detecção de anomalias em séries temporais multivariadas aplicadas a temporadas de furacões. In: V Seminário de Avaliação de Pesquisa Científica e Tecnológica [Internet]. Salvador, Bahia: SENAI CIMATEC; 2020. p. 3. Available from: https://www.researchgate.net/publication/343252454_ALGORITMO_DE_APRENDIZAGEM_DE_MAQUINA_NAO_SUPERVISIONADO_PARA_DETECCAO_DE_ANOMALIAS_EM_SERIES_TEMPORAIS_MULTIVARIADAS_APLICADAS_A_TEMPORADAS_DE_FURACOES

[6] Kim M, Ou E, Loh P-L, Allen T, Agasie R, Liu K. RNN-Based online anomaly detection in nuclear reactors for highly imbalanced datasets with uncertainty. Nucl Eng Des [Internet]. 2020 Aug;364(April):110699. Available from: https://doi.org/10.1016/j.nucengdes.2020.110699

[7] Schwabacher M, Oza N, Matthews B. Unsupervised Anomaly Detection for Liquid-Fueled Rocket Propulsion Health Monitoring. J Aerosp Comput Information, Commun [Internet]. 2009 Jul;6(7):464-82. Available from: https://arc.aiaa.org/doi/10.2514/1.42783

[8] Elavarasan D, Vincent DR, Sharma V, Zomaya AY, Srinivasan K. Forecasting yield by integrating agrarian factors and machine learning models: A survey. Comput Electron Agric [Internet]. 2018;155(October):257-82. Available from: https://doi.org/10.1016/j.compag.2018.10.024

[9] Khan S, Liew CF, Yairi T, McWilliam R. Unsupervised anomaly detection in unmanned aerial vehicles. Applied Soft Computing [Internet]. 2019 Oct;83:105650. Available from: https://doi.org/10.1016/j.asoc.2019.105650

[10] Ward CP, Weston PF, Stewart EJC, Li H, Goodall RM, Roberts C, et al.

Condition monitoring opportunities using vehicle-based sensors. In: Proceedings of the Institution of Mechanical Engineers, Part F: Journal of Rail and Rapid Transit. 2011.

[11] Yin S, Ding SX, Xie X, Luo H. A review on basic data-driven approaches for industrial process monitoring. IEEE Transactions on Industrial Electronics. 2014.

[12] Yiakopoulos CT, Gryllias KC, Antoniadis IA. Rolling element bearing fault detection in industrial environments based on a K-means clustering approach. Expert Systems with Applications [Internet]. 2011 Mar;38(3):2888-911. Available from: http://dx.doi.org/10.1016/j.eswa.2010.08.083

[13] Wang X-B, Zhang X, Li Z, Wu J. Ensemble extreme learning machines for compound-fault diagnosis of rotating machinery. Knowledge-Based Syst [Internet]. 2019; Available from: https://www.scopus.com/inward/record.uri?eid=2-s2.0-85071672820&doi=10.1016%2Fj.knosys.2019.105012&partnerID=40&md5=a47687f0f999c07d4b3a5a2ed5a8eb2b

[14] Nguyen VH, Cheng JS, Yu Y, Thai VT. An architecture of deep learning network based on ensemble empirical mode decomposition in precise identification of bearing vibration signal. Journal of Mechanical Science and Technology 2019;33(1):41-50.

[15] Abid A, Khan MT, Khan MS. Multidomain Features-Based GA Optimized Artificial Immune System for Bearing Fault Detection. IEEE Trans Syst Man, Cybern Syst [Internet]. 2020 Jan;50(1):348-59. Available from: https://ieeexplore.ieee.org/document/8031077/

[16] Lin J, Keogh E, Lonardi S, Chiu B. A symbolic representation of time series, with implications for streaming algorithms. In: Proceedings of the 8th ACM SIGMOD workshop on Research issues in data mining and knowledge discovery - DMKD '03 [Internet]. New York, New York, USA: ACM Press; 2003. p. 2-11. Available from: http://portal.acm.org/citation.cfm?doid=882082.882086

[17] Mohammad Y, Nishida T. Robust learning from demonstrations using multidimensional SAX. In: 14th International Conference on Control, Automation and Systems (ICCAS) [Internet]. Gyeonggi: IEEE; 2014. p. 64-71. Available from: http://ieeexplore.ieee.org/document/6987960/

[18] Cover T, Hart P. Nearest neighbor pattern classification. IEEE Trans Inf Theory [Internet]. 1967 Jan;13(1):21-7. Available from: http://ieeexplore.ieee.org/document/1053964/

[19] Gou J, Ma H, Ou W, Zeng S, Rao Y, Yang H. A generalized mean distance-based k-nearest neighbor classifier. Expert Syst Appl [Internet]. 2019 Jan;115:356-72. Available from: https://linkinghub.elsevier.com/retrieve/pii/S0957417418305293

[20] Efron B. Bootstrap Methods: Another Look at the Jackknife. In: Breakthroughs in Statistics [Internet]. 1992. p. 569-93. Available from: http://link.springer.com/10.1007/978-1-4612-4380-9_41

[21] Efron B, Rogosa D, Tibshirani R. Resampling Methods of Estimation. In: Wright JD, editor. International Encyclopedia of the Social & Behavioral Sciences [Internet]. Second Edi. Oxford: Elsevier; 2015. p. 492-5. Available from: http://www.sciencedirect.com/science/article/pii/B9780080970868421653

[22] Guha S, Mishra N, Roy G, Schrijvers O. Robust random cut forest based anomaly detection on streams. In: 33rd International Conference

on Machine Learning, ICML 2016 [Internet]. 2016. p. 3987-99. Available from: http://proceedings.mlr.press/v48/guha16.pdf

[23] National Hurricane Center Administration. Tropical Depression Nine Public Advisory One [Internet]. NINTH DEPRESSION OF THE SEASON FORMS EAST OF THE LESSER ANTILLES. Miami; 2012. Available from: https://www.nhc.noaa.gov/archive/2012/al09/al092012.public.001.shtml

[24] Avila L. Hurricane Rafael Tropical Cyclone Report [Internet]. Miami; 2013. Available from: http://www.nhc.noaa.gov/data/tcr/AL172012_Rafael.pdf

[25] Blake ES, Kimberlain TB, Berg RJ, Cangia, Losi JP, Beven II JL. Tropical cyclone report Hurricane Sandy [Internet]. National Weather Service, National Hurricane Center. Miami; 2013. Available from: https://www.nhc.noaa.gov/data/tcr/AL182012_Sandy.pdf

Chapter 5

Therapeutic Effect of Infra-Low-Frequency Neurofeedback Training on Children and Adolescents with ADHD

Horst Schneider, Jennifer Riederle and Sigrid Seuss

Abstract

In this observational study the outcomes of an EEG-based infra-low-frequency (ILF) neurofeedback intervention on patients with attention deficit (hyperactivity) disorder (ADHD) are presented. The question is addressed whether this computer-aided treatment, which uses a brain-computer-interface to alleviate the clinical symptoms of mental disorders, is an effective non-pharmaceutical therapy for ADHD in childhood and adolescence. In a period of about 15 weeks 196 ADHD patients were treated with about 30 sessions of ILF neurofeedback in an ambulant setting. Besides regular evaluation of the severity of clinical symptoms, a continuous performance test (CPT) for parameters of attention and impulse control was conducted before and after the neurofeedback treatment. During and after the therapy, the patients did not only experience a substantial reduction in the severity of their ADHD-typical clinical symptoms, but also their performance in a continuous test procedure was significantly improved for all examined parameters of attention and impulse control, like response time, variability of reaction time, omission errors and commission errors. In a post neurofeedback intervention assessment 97% of patients reported improvement in symptoms of inattention, hyperactivity or impulsivity. Only 3% of the patients claimed no noticeable alleviation of ADHD-related symptoms. These results suggest that ILF neurofeedback is a clinically effective method that can be considered as a treatment option for ADHD and might help reducing or even avoiding psychotropic medication.

Keywords: Infra-low frequency (ILF) neurofeedback, ADHD, therapy, continuous performance test, clinical study

1. Introduction

Hyperkinetic disorder, also known as attention deficit disorder (ADD) or attention deficit hyperactivity disorder, is a disorder that typically occurs in childhood. The core symptoms include increased inattention and/or hyperactivity and impulsivity as well as lack of emotional self-control and motivation. ADHD is a complex psychiatric and neurologically based disorder that usually is comorbid with other conditions: over one-half of children with ADHD have accessory symptoms like learning disabilities, conduct disorders, poor coordination, depression, anxiety, obsessive–compulsive disorders and bipolar disorders [1, 2]. Accordingly, the pathophysiological causes of ADHD are to be found in the central nervous system (CNS). Corresponding studies on ADHD patients show changes in dopaminergic and noradrenergic neurotransmission [3–6] as well as a (presumably related) developmental delay of the cortex, especially in the prefrontal region relevant for executive functions, attention and motor control [7]. In addition to these functional changes in defined brain areas, functional imaging studies in ADHD patients also have demonstrated changes in neuronal networks, e.g., in frontostriatal, fronto-parietal and ventral attention networks [8, 9] and in the default mode network (DMN) [10].

According to current estimations about five percent of children worldwide meet the diagnostic criteria of ADHD [11] and if left untreated, symptoms may persist into adulthood. Therefore, innovative and effective treatment methods that show long-lasting effectivity without the accompanying unwanted side effects of psychotropic drugs are of great relevance. Neurofeedback has been proven to be a treatment method that offers comparable effects in the therapy of ADHD like the use of pharmacological substances such as methylphenidate [12–17]. Follow-up studies and meta-analyses six, 12 or even 24 months after neurofeedback treatment show a sustained improvement of ADHD core symptoms [18, 19].

Neurofeedback is a computer-aided therapy method for clinical use, mainly as a treatment for mental disorders with the aim to improve self-regulation processes of the brain using a brain-computer interface (BCI). During a neurofeedback session selected parameters of the patient's electroencephalogram (EEG) are extracted according to their frequency and power density, processed, transformed into audio-visual feedback signals which then are being made perceptible for the patient's sensory organs by computer animations. By utilizing specific frequency components of the continuously measured full band EEG, the corresponding cerebral activities and their dynamics are reported back (feedback) to the central nervous system from where they originate. Due to the high performance of today's modern EEG and computer systems, electrical potential fluctuations of cerebral origin can continously be recorded from the skull with a high dynamic range. Furthermore, the neurofeedback-specific processing up to the generation and visual and acoustic presentation of the feedback signals can take place almost in real time, so that there is a minimal time delay only between the brain's generation of electrical activity, its electroencephalographical measurement and the presentation and perception of the EEG-derived audio-visual feedback signals. As a result, the brain can interact with the perceptual audio-visual "echo" of parts of its own activity, by improvement of its self-regulatory abilities [20, 21].

It has been known for a long time that brain functions can be influenced by feedback mechanisms [22], but neurofeedback was only developed in the late 1960s – without its clinical potential being recognized at first. A few years later, the first clinical studies showed particularly good therapeutic success using this technique in patients with severe epilepsy [23–26]. It was later shown that the effects remained even ten years after the end of the neurofeedback treatment [27]. Since self-regulation is an essential and fundamental function of the brain, the clinical treatment spectrum of neurofeedback is broad. Thus, in addition to epilepsy and the already mentioned hyperkinetic disorder, neurofeedback has also been shown to be an appropriate treatment for many other neurological disorders involving brain dysregulation, such as autism spectrum disorder (ASD) [28–33], migraine [34, 35], post-traumatic stress disorder (PTSD) [36–40], schizophrenia [20] and several others.

The various neurofeedback methods used typically differ in the extraction of the frequency components of the measured EEG that are used to calculate and control the feedback signals. In so-called frequency band training, the focus is on conventional frequency ranges of the human EEG between 1 and 40 Hz. Brain activities in this range usually dominate the EEG due to their clearly visible wave-like characters. It has long been confirmed in clinical studies that neurofeedback training in these frequency ranges, namely 4–8 Hz (theta range), 12–15 Hz (sensorimotor rhythm, SMR), and 16–20 Hz (beta range), can be an appropriate and effective treatment for children with ADHD [40–42]. However, the full band EEG also contains long-lasting potential shifts that are assigned to slow activities of the frequency range below 0.1 Hz. Such potential fluctuations typically are created by cortical neurons in preparation for sensomotoric tasks as well as for motor or cognitive behavior and events [16, 43]. According to their functional significance, these voltage signals are either classified as readiness potentials or, according to their time course, referred as slow cortical potentials (SCPs). It is assumed that slow surface negative potentials of cortical neurons represent a measure for the excitability of cortical neurons, while positive defections of such SCPs in the EEG signify a widespread absence of facilitation [43–45]. By influencing SCPs with weak external direct current voltage stimuli applied to the head, it could be shown that slow cortical negativity in certain cortical areas leads to better performance in sensorimotor tasks [16]. Abnormalities in SCP size seem to affect behavior and it has, for instance, been shown that children with ADHD show EEG abnormalities in the frequency range of SCPs [46, 47]. Children with attention deficits show smaller negative SCPs during the anticipation phase of a task in comparison to children without attention problems [16]. The two neurofeedback training methods that utilize such slow potentials in the EEG are ILF- and SCP-neurofeedback. Various studies document SCP neurofeedback training as an effective form of therapy for ADHD [18, 48, 49].

ILF neurofeedback was primarily developed empirically based on clinical observations from the frequency band and SCP methods. It utilizes the conventional frequencies between 1 and 40 Hz within nine fixed bands and transforms any dynamic progression of their spectral power above individual thresholds into a certain set of feedback signals ("Inhibits"). By this mechanism, the brain receives feedback about sudden changes in spectral power densities, which are linked directly to the respective brain activity components in the EEG. At the same time, the amplitudes and dynamics of the very slow cortical potentials of the "infra-low" frequency range of <0.1 Hz are determined in the EEG and, after setting an individual gain factor via a lowpass filter cutoff frequency by the therapist, transferred as a second set of feedback signals ("Signal"). The ILF neurofeedback protocol determines that the EEG is recorded in a bipolar montage. Thus, not the dynamically changing brain activity underneath each two electrodes is the targeted signal but their ratio and consequently, ILF neurofeedback represents a coherence training.

Other essential and standalone elements of the ILF neurofeedback protocol are that neither specific frequencies of brain activity in the EEG are actively promoted or suppressed via the feedback process, nor is the patient supposed to produce brain activity of specific frequencies voluntarily. Rather, the therapeutic work in ILF neurofeedback is based on the assumption that the symptoms of the patient indicate over- or under-excitation in certain multiple association areas of the brain [50]. By placing the EEG electrodes above such multiple association areas on the head of the patient, the brain receives continuous feedback on its internal states. This happens via up to 15 different computer-generated audio-visual feedback signal parameters to trigger neurophysiological modulation on an unconscious level.

The patient may become aware of the feedback-induced cerebral changes through the conscious perception of temporary positive sensations, like relaxation, increased concentration or motoric calmness or a reduced level of alertness. However, such temporary sensations could also be mild sensations of fatigue, headaches, increased motor activity or dizziness and thus, unwanted effects. The therapist is therefore encouraged, to always observe the patient for signs of relaxation, stress, comfort or discomfort and to also inquire at regular intervals about perceived feelings. In case of positive observations or reports from the patient the therapist will proceed with the actual settings of the training parameters or change them to eliminate unwanted effects.

In addition to these partly subjective effects of the training, there were recently also reports published that demonstrate defined neurophysiological changes in the brain which can be attributed to the use of ILF neurofeedback. A quantitative analysis of 19-channel EEG recordings before and after 20 sessions of ILF neurofeedback training shows a significant increase in spectral power in the 0.5 Hz frequency band [51, 52]. The general increase in spectral power of the ILF component of the EEG indicates that ILF neurofeedback training induces a modified baseline brain state. Another study using functional magnetic resonance imaging (fMRI) shows that even a single session of ILF neurofeedback leads to significant changes in connectivity in the brain [53].

While SCP and frequency band training have been used for many years to treat ADHD, there are only a few studies in which ILF neurofeedback has been used as a treatment method [54]. ILF neurofeedback could represent a particularly effective treatment method for pathologies in which the brain is dysregulated. It combines the above-mentioned components that characterize the procedure with the methodological immanence for the therapist to adapt the treatment to the patient's individual symptomatology. In consequence, the natural question arises concerning the evidence-based level of ILF neurofeedback therapy. The present study therefore aims to clarify the question whether ILF neurofeedback is an effective therapy for children and adolescents with ADHD. In addition, little research has been done on the effectiveness of neurofeedback for ADHD in everyday life, so the present study tracks the individual symptom profiles. This examines if the effect of ILF neurofeedback leads to an improvement in life quality of those affected.

2. Methods

2.1 Study operator and therapists

The present study was conducted as a pilot project of a network of five practices for child and adolescent psychiatry in Germany. It is in accordance with the declaration of Helsinki. All interventions mentioned in this study were carried out by a total of 25 specialist therapists who had qualified in a certified training course of ILF neurofeedback lasting several days.

The data of the present observational study was collected by the participating practices in the course of treatment of their patients. A declaration of consent for the anonymized collection and processing of the data in the sense of an observational study with a pilot character was enclosed with the treatment contract, which the patients received before the start of the therapy, and which was signed by the patients or, in the case of minors, by the parents.

2.2 Participants

Participants in this study were recruited from children and adolescents who visited one of the participating practices due to ADHD-related symptoms or already diagnosed ADHD. Some of them were already under drug treatment with methylphenidate medication at the beginning of the study. In addition to age and informed consent, the clinically validated diagnosis of attention deficit (hyperactivity) disorder was another inclusion criterion for study participation. **Figure 1** shows the inclusion criteria for the study.

A total of 251 patients participated in the data collection and received therapy in the form of ILF neurofeedback treatment. On average, these patients had an age of 12.1 years (SD: 2.8, interval: 7.3–21.5), with 82% belonging to the age group 7–14 years and 18% to the age group 15–21 years. The gender distribution of the participants shows a majority of 79% males and 21% females.

2.3 Study design

In the present observational study, a symptom tracking procedure was used to measure subjectively perceived expression and severity of ADHD-typical symptoms before the start (T_0) and at the end (T_2) of ILF neurofeedback therapy. A QIKtest device was used for continuous performance tests to measure for attention, sustained attention, and impulse control at T_0 and T_2. For each participant the therapy consisted of approximately 30 ILF neurofeedback sessions, each lasting up to 50 minutes, with about two sessions every week and thus, a therapy period of about 15 weeks (see **Figure 2**).

2.4 Electrophysiology and software

The ILF neurofeedback interventions were performed with neurofeedback systems from BEE Medic Inc. (Germany), that consist of a 2-channel EEG differential amplifier EEG NeuroAmp® II (Corscience Inc., Germany) with full bandwidth (DC to 100 Hz), 32 bit resolution, a sampling rate of 500 sps and integrated impedance meter (impedance range 0–140 kOhm) as well as the software Cygnet® (BEE Medic Inc., Germany).

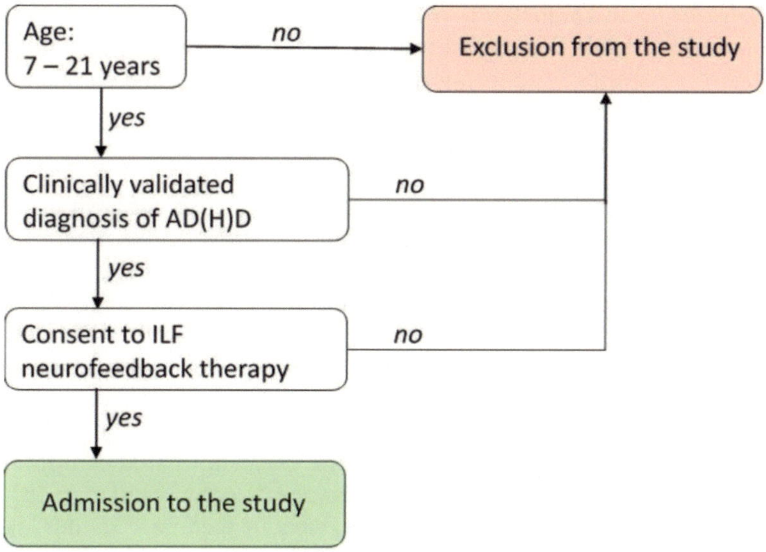

Figure 1.
Inclusion and exclusion criteria of this observational study. Included into the study were children and adolescents with an age between 7 and 21 years, a diagnosed hyperkinetic disorder and a signed consent for a ILF neurofeedback therapy.

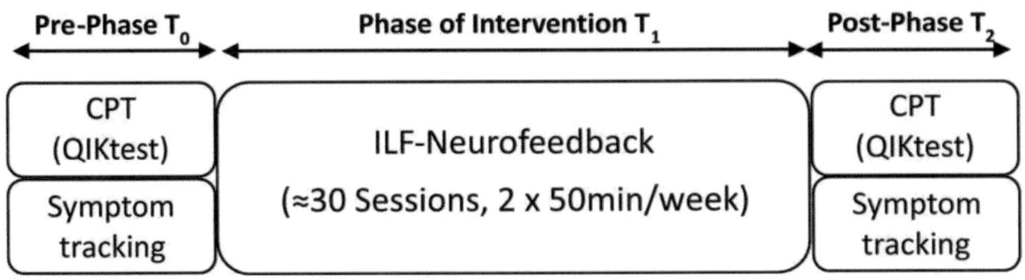

Figure 2.
Study design showing the different phases.

Before applying the electrodes, the skin at the electrode positions was treated with an abrasive cleaning paste (Nuprep®, Weaver and Company, USA) and then Ag/AgCl electrodes were applied using conductive paste (Ten20®; Weaver and Company, USA) to ensure a proper conductivity.

2.5 ILF neurofeedback protocol

The used neurofeedback-method was according to the ILF neurofeedback protocol and followed the description of Susan Othmer [55]. It consists of a 2-channel EEG that was recorded from the scalp of a patient using a bipolar montage and electrode placement sites in accordance with the international 10–20 EEG system. Electrodes were placed individually according to the protocol guide [55], with starting placements at T3-T4 or T4-P4 electrode sites.

The neurofeedback process and the audio visual feedback was controlled and applied using Cygnet® software. During continuous EEG recording, features of the EEG were extracted in near real time to build two different dynamically changing components of the feedback process: "Inhibits" and "Signal".

To calculate the "Inhibits" component, the supra-threshold EEG power densities of nine filter blocks in fixed frequency steps in the range between 1 and 40 Hz were summed up. The thresholds of the nine frequency bands were individually and dynamically set and adjusted to maintain the actual EEG power density of a frequency band to be sub-threshold for about 95% of the time. Due to this calculation method of dynamically adapting threshold values, a sudden increase in power density in an EEG frequency band instantly leads to suprathreshold values and thus immediately to an increase in the "inhibits" component.

To calculate for the "Signal" component the EEG power density of a "infra-slow" frequency band was extracted and determined. In the neurofeedback protocol used "infra-slow" frequencies are defined as frequencies below 0.1 Hz. Accordingly, the therapist is required to set the cut-off frequency of a low-pass filter in the millihertz frequency range via the software in order to extract the "infra-slow" "signal" component from the EEG and to continuously determine its signal strength.

One of the core features of the ILF neurofeedback is the subsequent transformation of the continuously determined "inhibit" as well as "signal" components into animated audio-visual feedback signals, which are presented to the patient on a separate computer screen. Typically, this is done via an animated computer game in which certain acoustic and visual parameters are directly coupled to either the "inhibit" or "signal" component or their ratio. Various feedback "games" were available to the ADHD patients for free selection and their common feature was that the calculated "inhibit" component modulated the volume of the underlying music and determined the color contrast and brightness of the animated environment. The simultaneous modulatory effects of the "Signal" component concerned the speed of the animated game character and the volume of its sounds.

The promotion of CNS stability is the first objective of brain training [37]. Because brain stability is an individual feature, an individualized training strategy, in which the reinforcement "infra-slow" frequency is optimized for each individual, is a mandatory element of the ILF neurofeedback protocol [55]. According to the protocol, the "signal" frequency has to be adjusted by the therapist during the first sessions to the state in which the person is maximally calm, attentive and as euthymic as the nervous system is capable of being at that moment. The fine-tuning of the optimal reinforcement frequency (ORF) then is done on the basis of reports from the patient on their own status or observations of the therapist. In this study, the ORF for the infra-low signal was determined individually during the first 1–3 sessions based on the report of the patient or from observing behavioral signs of stress, alertness, wellbeing or relaxation on the patient by the therapist. Thereafter, the ILF neurofeedback therapy was proceeded with the "signal" frequency set to the patient's individual ORF.

2.6 Continuous performance test (CPT)

In order to measure changes in attention, sustained attention and impulse control, a CPT with the QIKtest device (BEE Medic Inc., Germany) was carried out before the start and at the end of neurofeedback therapy. The QIKtest is a mobile, stand-alone test display/input device with a standardized test procedure that is used in particular to record selective attention, sustained attention and impulsive behavior. The CPT of the QIKtests consist of displaying "GO/NO GO "tasks for 21 minutes. The test is divided into five phases, in which the occurrence, incidence and intervals of "GO" tasks differ to measure four parameters of attention: average reaction time (RT), variability of reaction time (VAR), omission errors (OM) and commission errors (CO).

During the CPT, two simple visual conditions ("target"/"GO" and "non-target"/"NOGO") are presented once every two seconds to the patients on the screen of the QIKtest device via nine luminous fields: "GO" when all fields except the middle field light up and "NO-GO" when all nine fields light up.

In a period of 2 seconds, in a seemingly (for the patient) random fashion one of the two stimulus conditions lights up for a duration of 100 milliseconds. The subject's task is to press a button on the QIKtest device as quickly as possible only when the "GO" condition appears. This results in two possible types of errors: Omission errors, when the required reaction to the "GO" condition failed to appear, and commission errors, when the reaction button on the QIKtest device was pressed after a "NO-GO" signal was displayed. In addition, the QIKtest device measures the reaction time for each correct reaction with a measurement accuracy of 0.1 milliseconds and calculates RT and VAR.

The statistical evaluation of the test results was carried out using PSPP (GNU project, open source), version 1.2.0.

In order to qualitatively classify changes in the investigated attention parameters and those of impulse control, the CPT database of EEG Expert (EEG Expert Limited, Ankara, Turkey) was used. The "equivalent mental age", derived from the mean result of a reference group for the specific age, was determined for RT, VAR, OM and CO from the corresponding norm curves. The CPT database contains >50,000 records of individuals of both sexes aged 6–70 years in 40 age groups, with at least 500 records per age group.

2.7 Symptom tracking

To assess symptom changes through ILF neurofeedback therapy, patients were asked to track their individual symptoms out of a catalog of 137 ADHD-specific and other symptoms from the categories of sleep, attention and learning behavior, sensory and perception, behavior, emotions, physical symptoms and pain, before (T_0) and after the ILF neurofeedback intervention (T_2). Between the two points of measurement (T_0 = Pre and T_2 = post) was the phase of neurofeedback intervention (T_1) (see **Figure 2**). Participating patients could indicate a severity level between 0 (symptom does not apply at all) and 10 (symptom occurs very frequently or is maximal pronounced) for each of the 137 given symptoms.

The statistical evaluation of the symptom survey was done using the software PSPP, version 1.2.0.

3. Results

Of the 251 ADHD patients treated with ILF neurofeedback during the entire data collection period, only 196 had pre-post QIKtest data collected. The average duration of therapy in terms of neurofeedback sessions was 38.5 (SD = 21.6), three participants dropped out of therapy.

3.1 Continuous performance test

The pre-post data at T_0 and T_2 of 196 participants were included in the evaluation of the continuous performance test using the QIKtest device. Changes in four variables were analyzed: average reaction time (RT), variability of reaction time (VAR), omission errors (OM) and commission errors (CO). The averaged RT of the patients improved during the duration of the ILF neurofeedback training by about 21 ms - from 457 ms at T_0 to 436 ms at T_2 (see **Table 1**). In parallel, VAR improved as well by about 18 ms - from 122 ms at T_0 to 104 ms at T_2. To examine their statistical significance, the values of RT and VAR were compared separately using independent Student's *t*-tests, as a normal distribution with equal variances was given. According to the t-test results, the improvements of RT and VAR after ILF neurofeedback treatment were statistically highly significant (see **Table 1**). The third attention parameter that was measured, OM, too improved from an average of 9.6 errors (SD = 15.1 errors) at T_0 to 5.0 errors (SD = 9.3 errors) at T_2. The test parameter that determines impulse control CO improved from 19.1 errors (SD = 17.3 errors) on average at T_0 to 9.0 errors (SD = 9.0 errors) at T_2. The significance of the improvements was examined statistically using a non-parametric Wilcoxon signed-rank test, because OM and CO did not follow a normal distribution. According to their Wilcoxon signed rank test results, the improvements of OM and CO after ILF neurofeedback treatment were statistically highly significant (see **Table 1**).

To investigate the relevance ("quality") of the improvements in the studied parameters of attention and impulse control in relation to mental maturity, the respective "equivalent mental age" for RT, VAR, OM and CO was determined from

N = 196	Pre (T_0)	Post (T_2)	Difference	p
Reaction Time (RT)	457 ± 88 ms	436 ± 85 ms	−21 ms	<0.0001[1]
Variability of RT (VAR)	122 ± 31 ms	104 ± 30 ms	−18 ms	<0.0001[1]
Omission Errors (OM)	9.6 ± 15.1	5.0 ± 9.3	−4.6	<0.0001[2]
Commission Errors (CO)	19.1 ± 17.2	9.0 ± 9.0	−10.1	<0.0001[2]

[1]*Student's t-test.*
[2]*Wilcoxon signed rank test.*

Table 1.
Results of the continuous performance test.

the corresponding norm curves of the CPT database. On average, the participating ADHD patients had an age of 12.1 years. However, their averaged performance in the CPT before the start of the ILF neurofeedback training was clearly below their averaged actual age when compared with the CPT database (see **Figure 3**): the averaged performances for the attention parameters RT, VAR and OM of the average 12.1-year-old ADHD patients corresponded to the 10.2 (RT), 10.0 (VAR) and 8.9 (OM) years age groups in the CPT database and thus, lack a mental maturity of around 2 years. For the tested parameter of impulse control, CO, the averaged performances of the ADHD patients corresponded to the 8.5 (CO) years age group in the CPT database and thus, showed an even slightly more delayed mental maturity of about 3.5 years.

In terms of "equivalent mental age" that was derived from the CPT database, the ADHD patients benefited considerably from the therapy. After the ILF neurofeedback training equivalent mental age of the ADHD patients clearly increased for RT from 10.2 to 12.3 years, for VAR from 10.0 to 12.8 years, for OM from 8.9 to 10.3 years and for CO from 8.5 to 15.0 years (see **Figure 3**).

3.2 Symptom tracking

According to the patients' self-disclosure or evaluation by the therapists, 97% of the patients experienced an improvement of the symptoms which had been individually perceived as stressful before the neurofeedback therapy, like inattention, hyperactivity, impulsivity, difficulties to fall asleep, distractibility, rage and others. Only 3% of the patients claimed no noticeable improvement of the symptoms.

The course of symptom severity before, during and after approximately 30 sessions of ILF neurofeedback was assessed in 43 ADHD patients for the three core symptoms of their disorder: inattention, hyperactivity and impulsivity. Before the start of the ILF neurofeedback intervention at T_0 the patients evaluated the core symptoms of their disorder as to be very pronounced, with high average values for inattention, hyperactivity and impulsivity (see **Table 2**).

Comparison of these averaged severity values with the individually evaluated severity levels of these symptoms after the treatment with ILF neurofeedback at T_2 and determination of the level of significance by Wilcoxon signed rank test, all three core symptoms had been improved significantly. For the symptom of inattention, the participants reported at T_2 a highly significant decrease of the individually perceived severity by 2.9, for hyperactivity by 3.5 and impulsivity by 1.3 (see **Figure 4**).

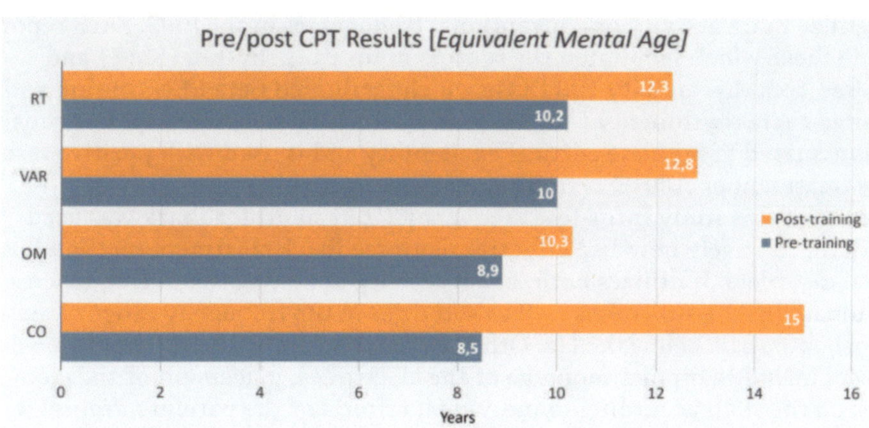

Figure 3.
Improvements of the equivalent mental age for the different test parameters of the continuous performance test.

Symptom	Time	Severity of Symptoms	p
Inattention	T_2	5.4 ± 2.2	0*
	T_0	8.3 ± 1.4	
Hyperactivity	T_2	4.6 ± 2.5	0.012*
	T_0	8.1 ± 1.2	
Impulsivity	T_2	6.1 ± 2.5	0.018*
	T_0	7.4 ± 1.8	

Table 2.
Severity of the different ADHD symptoms (statistics by Wilcoxon signed rank test).= statistical significance attained.*

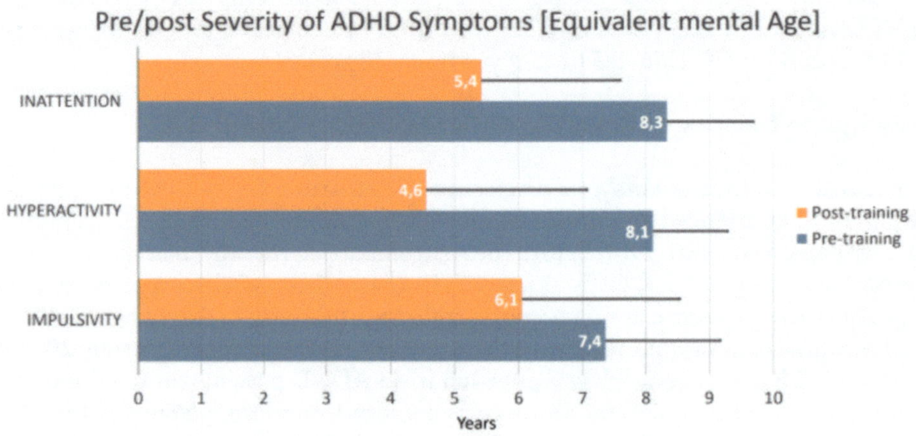

Figure 4.
Improvements in ADHD symptom ratings.

4. Discussion

Since the first reports of successful neurofeedback treatment in ADHD [56], several studies have investigated the effects on symptoms of ADHD such as inattention, impulsivity and hyperactivity with neurofeedback protocols that utilize brain activity of conventional frequencies in the EEG. Such reports include those which facilitated the sensorimotor EEG rhythm (SMR) and inhibited beta rhythmicity and those which facilitated beta EEG rhythm and inhibited theta rhythmicity [40, 42, 57–60]. Another neurofeedback approach that is assumed to regulate cortical excitability and is used with positive results in the treatment of ADHD is training of Slow Cortical Potentials (SCP) [48, 61]. However, in this study Infra-low Frequency (ILF) neurofeedback was used, a modern, relatively new and effective neurofeedback treatment method for mental disorders. It utilizes both, brain activity of conventional frequencies in the human EEG (1–40 Hz) as well as activities in the frequency range of slow cortical potentials below 0.1 Hz. Other characteristics of the ILF neurofeedback protocol include a bipolar montage of the electrodes, placement of the electrodes on the skull according to individual criteria of the patient's arousal level and mental strength, and continuous feedback of the parameters extracted from the full-band EEG in audio-visual computer animations that have a game-like character.

Recent reports demonstrate that ILF neurofeedback not only utilizes slow brain activity in the EEG but also can directly lead to a significant increase in spectral power in the sub 0.5 Hz frequency band [51, 52]. Clinically, it has been shown that children with attention deficits show smaller negative SCPs during the anticipation phase of a task in comparison to children without attention problems [16] or other EEG abnormalities in the frequency range of SCPs [46, 47]. In the light of these findings, we conducted this multi-center study to address the question of whether ILF neurofeedback is an effective and significant treatment for ADHD and leads to an improvement in quality of life of those affected.

A total of 251 ADHD child and adolescent patients were included in this study and received a treatment consisting of an average of 39 ILF neurofeedback sessions over a period of at least 15 weeks (about two sessions of neurofeedback per week). Only three patients decided to discontinue treatment prematurely. Although we did not investigate this aspect scientifically, it can be concluded from the low dropout rate that the ILF neurofeedback was well accepted as a treatment method by the vast majority of the ADHD patients (and their parents). According to the patients' self-disclosure or evaluation by the therapists, 97% of the patients reported an improvement of the symptoms which had been individually perceived as stressful before the neurofeedback therapy. Only 3% of the patients claimed no noticeable improvement of the symptoms by the ILF neurofeedback training. The general effect of the ILF neurofeedback treatment therefore can be rated as excellent.

In order to make the patients' subjective assessment of their symptoms measurable, they were asked before and after the end of treatment to perform an evaluation of their most prominent symptoms on the basis of severity levels between 0 and 10. The most severe symptoms were chosen from a questionnaire of 137 ADHD-specific and other symptoms. This included the categories sleep, attention and learning behavior, sensory and perception, behavior, emotions, physical symptoms and pain. Regarding symptom tracking, complete data sets were unfortunately only available from 43 patients (and thus only from about 1/6 of the participating children and adolescents). Nevertheless, the size of this sample is sufficient for a statistical analysis in which we focused on the three core symptoms of the ADH disorder, inattention, hyperactivity and impulsivity. Before the ILF neurofeedback intervention, the severity of inattention was rated to be at 8.3 in average and thus, experienced as to be very pronounced. A similar average severity level was reported by the participants for the symptom of hyperactivity, which was 8.1. The impulsivity was rated at 7.4 on average and thus, only slightly less severe than the aforementioned symptoms. This shows that the three core symptoms of ADH disorder are indeed perceived by the patients as highly burdening. After the therapy of approx. 30 sessions of ILF neurofeedback, the patients assessed these symptoms as significantly less stressful, with a clear average improvement in inattention by 1.9 severity points and in hyperactivity by as much as 3.5 severity points. Regarding the severity of their impulsivity, the participating children and adolescents rated slight but significant decrease of 1.3 severity points after the treatment. From these results, it can be concluded that 30 sessions of ILF neurofeedback, according to the subjective perception of the patients, are sufficient to improve hyperactivity and inattention symptoms in children and adolescents with ADHD. The treatment can also lead to a slightly milder, but still significant improvement in impulsivity in the same group of patients. These effects of ILF neurofeedback therapy are in accordance with the results of controlled studies on ADHD using other neurofeedback protocols. In these studies high to moderate effect sizes were also found on inattention and impulsivity as well as on hyperactivity ([12, 13, 15, 62, 63], for a review see [64]).

These positive results are mainly based on the subjective sensations and experiences of ADHD patients. In order to examine and monitor the quality and effectiveness of the ILF neurofeedback treatment on the basis of more objective criteria, the participants completed a 21-minute visual GO/NOGO continuous performance test (CPT) before the start and after the end of the intervention. Through this measure the parameters of attention and impulse control could be directly examined in detail. The three attention parameters that were tested are the response time, the variability of the response time and omission errors. The reaction or response time (RT) is the mean of all correct reaction times to a target stimulus ("GO" condition) and is a measure of the speed of responses. This attention parameter is accompanied by the variability of the response time (VAR), which is a measure of the consistency of the response. Finally, omission errors occur when the subject does not respond correctly to a target stimulus, which is assessed as a sign of inattention. A comparison of the test results prior and after about 15 weeks of ILF neurofeedback intervention revealed a significant improvement of all three attention parameters. The averaged Reaction time decreased for 21 ms, VAR for 18 ms and the averaged OM by −4.6 errors. To transform these results into more tangible values, the conversion into an "equivalent mental age" (EMA) was done based on the large CPT database of EEG Expert. Here, the "equivalent mental age" indicates the specific age of the reference group whose norm test result corresponds with the test result of the patient.

The improvements in the three tested attention parameters are reflected in a significant increase in the EMA. Before the start of the ILF neurofeedback therapy the ADHD children and adolescents were about 2 years of EMA behind, but regarding the attention parameters examined, they were able to make up for this delay within the 15 weeks of neurofeedback training. Most prominent was the improvement in averaged consistency of the response time (VAR) which led to an increase of EMA by +2.8 years and the shorter mean response time (RT) which increased the EMA by +2.1 years. The improvement in omission errors was slightly less pronounced because it resulted to +1.4 years in equivalent mental age. For the three tested attention parameters it therefore can be stated that – within the 15 weeks period of ILF neurofeedback treatment - the brain of the ADHD patients had gained in maturation corresponding to a developmental progress of about two years.

Commission errors (CO) in the CP test occur when the patient responds (incorrectly) to a non-target ("NOGO") task, which makes this test parameter a good measure for impulsivity. In all participating patients, impulse control improved significantly from an average of 19.1 CO errors before the ILF neurofeedback treatment, to only 9.0 CO errors after the intervention. In terms of equivalent mental age, this means that the performance of the ADHD patients improved from a below-average of 8.5 years to an above-average EMA of 15.0 years after the EEG-assisted neurofeedback intervention.

All objective improvements in the attention and impulsivity parameters examined in the CP testing are completely consistent with the ADHD patients' subjectively perceived reductions in the severity of their symptoms of inattention, hyperactivity and impulsivity, which were rated as highly distressing prior to ILF neurofeedback treatment. Based on the data and feedback from clinicians and patients it therefore can be concluded that ILF neurofeedback can be seen as an effective method to treat ADHD in children and adolescents.

Due to the fact that ADHD on one hand is a complex psychiatric and neurologically based disorder which usually is associated with many comorbidities as social behaviors disorders, affective disorders, depression, anxiety, obsessive–compulsive disorders, bipolar disorders and others [1, 2] and ILF neurofeedback on the other hand is indicated for all of the mentioned ADHD comorbidities [65, 66]. It would therefore be interesting to undertake a more comprehensive evaluation of the symptom severities of ADHD patients and to investigate in a controlled study to what extent ILF neurofeedback therapy leads to further improvements in cerebral self-regulation, which also encompasses the areas of other comorbidities of ADHD.

5. Conclusion

This observational clinical study could show significant improvements in major symptoms of ADHD - being inattention, hyperactivity and impulsivity - along with an improvement of attention, sustained attention and impulse control as well as the mental age equivalents in young patients with ADHD after ILF neurofeedback intervention. These results fit in line with presented study outcomes on neurofeedback in the treatment of ADHD - given the particularity that symptom based and individualized ILF neurofeedback presents a modern approach to EEG neurofeedback therapy options. Patients, parents and therapists evaluated the implementation and therapeutic outcome pleasant and positive. Whatsoever based on this and prior results it can be concluded that neurofeedback can be assessed as an effective, non-invasive, non-drug and pain-free treatment opportunity enlarging the ADHD treatment options. These promising results should motivate further research, especially studies overcoming the limitations of this one and including an interventional design, control parameters, further validated research instruments and long-term observations.

From a therapeutic point of view ILF neurofeedback can add a value to the treatment of children and adolescents with ADHD but further and more controlled research is needed to determinate outcome differences, especially in comparison to standard of care treatment.

Author details

Horst Schneider*, Jennifer Riederle and Sigrid Seuss
BEE Medic GmbH, Singen, Germany

*Address all correspondence to: hschneider@beemedic.de

References

[1] Hickey G, Fricker P. Attention deficit hyperactivity disorder, CNS stimulants and sport. Sport Med. 1999;27(1):11-21.

[2] Prinz W. Neurofeedbacktherapie als Spezialtherapieangebot. Psychopraxis Neuropraxis. 2015;18(5):180-183.

[3] Shaw P, Gornick M, Lerch J, Addington A, Seal J, Greenstein D, et al. Polymorphisms of the dopamine D4 receptor, clinical outcome, and cortical structure in attention-deficit/ hyperactivity disorder. Arch Gen Psychiatry. 2007;64(8):921-931.

[4] Swanson JM, Kinsbourne M, Nigg JT, Lanphear B, Stefanatos GA, Volkow N, et al. Etiologic subtypes of attention-deficit/hyperactivity disorder: Brain imaging, molecular genetic and environment factors and the dopamine hypothesis. Neuropsychol Rev. 2007;17:39-59.

[5] Volkow ND, Wang GJ, Newcorn J, Telang F, Solanto M V., Fowler JS, et al. Depressed dopamine activity in caudate and preliminary evidence of limbic involvement in adults with attention-deficit/hyperactivity disorder. Arch Gen Psychiatry. 2007;64(8):932-940.

[6] Scassellati C, Bonvicini C. Role of Dopaminergic and Noradrenergic Systems as Potential Biomarkers in ADHD Diagnosis and Treatment. In: Norvilitis JM, editor. ADHD - New Directions in Diagnosis and Treatment. InTechOpen; 2015.

[7] Arnsten AFT. The Emerging Neurobiology of Attention Deficit Hyperactivity Disorder: The Key Role of the Prefrontal Association Cortex. J Pediatry. 2009;154(5).

[8] Bush G. Attention-deficit/ hyperactivity disorder and attention networks. Neuropsychopharmacology. 2010;35(1):278-300.

[9] Zepf FD, Bubenzer-Busch S, Runions KC, Rao P, Wong JWY, Mahfouda S, et al. Functional connectivity of the vigilant-attention network in children and adolescents with attention-deficit/hyperactivity disorder. Brain Cogn. 2019;131 (November 2017):56-65.

[10] Posner J, Park C, Wang Z. Connecting the dots: A review of resting connectivity MRI studies in attention-deficit/hyperactivity disorder. Neuropsychol Rev. 2014;24(1):3-15.

[11] Polanczyk G, De Lima MS, Horta BL, Biederman J, Rohde LA. The worldwide prevalence of ADHD: A systematic review and metaregression analysis. Am J Psychiatry. 2007;164(6):942-948.

[12] Fuchs T, Birbaumer N, Lutzenberger W, Gruzelier JH, Kaiser J. Neurofeedback treatment for attention-deficit/hyperactivity disorder in children: A comparison with methylphenidate. Appl Psychophysiol Biofeedback. 2003 Mar;28(1):1-12.

[13] Monastra VJ, Monastra DM, George S. The effects of stimulant therapy, EEG biofeedback, and parenting style on the primary symptoms of attention-deficit/hyperactivity disorder. Appl Psychophysiol Biofeedback. 2002;27(4):231-249.

[14] Rossiter T, La Vaque TJ. A comparison of EEG biofeedback and psychostimulants in treating attention deficit/hyperactivity disorders. J Neurother. 1995;1(1):48-59.

[15] Rossiter T. The effectiveness of neurofeedback and stimulant drugs in treating AD/HD: Part II. Replication. Appl Psychophysiol Biofeedback. 2004;29(4):233-243.

[16] Rockstroh B, Elbert T, Lutzenberger W, Birbaumer N.

Biofeedback: Evaluation and therapy in children with attentional dysfunctions. In: Rothenberger A, editor. Brain and Behavior in Child Psychiatry. Springer Berlin Heidelberg; 1990. p. 345-355.

[17] Lubar JF, Swartwood M, Swartwood J, Timmermann D. Quantitative EEG and auditory event-related potentials in the evaluation of attention-deficit/hyperactivity disorder: Effects of methylphenidate and implications for neurofeedback training. J Psychoeduc Assess. 1995;34(November):143-160.

[18] Gani C. Long term effects after feedback of slow cortical potentials and of theta/Beta - amplitudes in children with attention deficit hyperactivity disorder (ADHD). Medizinische Fakultät der Eberhard Karls Universität Tübingen. 2009.

[19] Van Doren J, Arns · Martijn, Hartmut Heinrich ·, Vollebregt MA, Strehl U, Loo SK. Sustained effects of neurofeedback in ADHD: a systematic review and meta-analysis. Eur Child Adolesc Psychiatry. 2019;28:293-305.

[20] Schneider F, Rockstroh B, Heimann H, Lutzenberger W, Mattes R, Elbert T, et al. Self-regulation of slow cortical potentials in psychiatric patients: Schizophrenia. Biofeedback Self Regul. 1992;17(4):277-292.

[21] Thibault RT, Lifshitz M, Raz A. The self-regulating brain and neurofeedback: Experimental science and clinical promise. Cortex. 2016;74:247-261.

[22] Jasper H, Shagass C. Conditioning of the occipital alpha rhythm in man. J Exp Psychol. 1941;28(5):373-388.

[23] Sterman MB, Macdonald LR, Stone RK. Biofeedback training of the Sensorimotor electroencephalogram rhythm in man: Effects on epilepsy. Epilepsia. 1974;15(3):395-416.

[24] Egner T, Sterman MB. Neurofeedback treatment of epilepsy: From basic rationale to practical application. Expert Rev Neurother. 2006;6(2):247-257.

[25] Sterman MB. Sensorimotor EEG Feedback Training in the Study and Treatment of Epilepsy. In: The Neurobehavioral Treatment of Epilepsy. 1993.

[26] Sterman MB, Egner T. Foundation and practice of neurofeedback for the treatment of epilepsy. Appl Psychophysiol Biofeedback. 2006;31 (1):21-35.

[27] Strehl U, Birkle SM, Wörz S, Kotchoubey B. Sustained reduction of seizures in patients with intractable epilepsy after self-regulation training of slow cortical potentials - 10 years after. Front Hum Neurosci. 2014;8(AUG):1-7.

[28] Kouijzer MEJ, de Moor JMH, Gerrits BJL, Congedo M, van Schie HT. Neurofeedback improves executive functioning in children with autism spectrum disorders. Res Autism Spectr Disord. 2009;3(1):145-162.

[29] Friedrich EVC, Sivanathan A, Lim T, Suttie N, Louchart S, Pillen S, et al. An effective neurofeedback intervention to improve social interactions in children with autism Spectrum disorder. J Autism Dev Disord. 2015;45(12):4084-4100.

[30] Kouijzer MEJ, de Moor JMH, Gerrits BJL, Buitelaar JK, van Schie HT. Long-term effects of neurofeedback treatment in autism. Res Autism Spectr Disord. 2009;3(2):496-501.

[31] Holtmann M, Steiner S, Hohmann S, Poustka L, Banaschewski T, Bölte S. Neurofeedback in autism spectrum disorders. Dev Med Child Neurol. 2011;53(11):986-993.

[32] Coben R, Padolsky I. Assessment-guided neurofeedback for autistic

spectrum disorder. J Neurother. 2007;11(1):5-23.

[33] Jarusiewicz B. Efficacy of neurofeedback for children in the autistic spectrum: A pilot study. J Neurother. 2002;6(4):39-49.

[34] Stokes DA, Lappin MS. Neurofeedback and biofeedback with 37 migraineurs: A clinical outcome study. Behav Brain Funct. 2010;6:1-10.

[35] Walker JE. QEEG-Guided Neurofeedback for Recurrent Migraine Headaches. Vol. 42, Clinical EEG and Neuroscience. 2011. p. 59-61.

[36] Nilsson RM, Nilsson V. Neurofeedback Treatment for Traumatized Refugees-a Pilot Study. 2014.

[37] Othmer S, Othmer SF, Legarda SB. Clinical neurofeedback: Training brain behavior. Treat Strateg Pediatr Neurol Psychiatry. 2011;2(1):67-73.

[38] Othmer S, Othmer SF. Post traumatic stress disorder—The neurofeedback remedy. Biofeedback. 2009;37(1):24-31.

[39] Peniston E, Kulkosky P. Alpha-Theta Brainwave Neuro-Feedback for Vietnam Veterans with Combat- Related Post-Traumatic Stress Disorder. Med Psyc~OIherapy. 1991;4:7-60.

[40] Scott WC, Kaiser D, Othmer S, Sideroff SI. Effects of an EEG biofeedback protocol on a mixed substance abusing population. Am J Drug Alcohol Abuse. 2005;31(3):455-469.

[41] Lubar JF. EEG biofeedback and learning disabilities. Theory Pract. 1985;24(2):106-111.

[42] Lubar JO, Lubar JF. Electroencephalographic biofeedback of SMR and beta for treatment of attention deficit disorders in a clinical setting. Biofeedback Self Regul. 1984;9(1):1-23.

[43] Birbaumer N, Elbert T, Canavan AGM, Rockstroh B. Slow Potentials of the Cerebral Cortex and Behavior. Physiol Rev. 1990;70(1).

[44] Elbert T. Slow potential changes in the human brain. In: McCallum WC, editor. Proceedings of a NATO Advanced Research Workshop on Slow Potential Changes in the Human Brain. Plenum Press; 1990. p. 235-251.

[45] Rockstroh B, Elbert T, Canavan AGM, Lutzenberger W, Birbaumer N. Slow Cortical Potentials and Behaviour. 2. Edition. Baltimore: Urban & Schwarzenberg; 1989.

[46] Perchet C, Revol O, Fourneret P, Mauguière F, Garcia-Larrea L. Attention shifts and anticipatory mechanisms in hyperactive children: An ERP study using the Posner paradigm. Biol Psychiatry. 2001;50(1):44-57.

[47] Dumais-Huber C, Rothenberger A. Psychophysiological correlates of orienting, anticipation and contingency changes in children with psychiatric disorders. J Psychophysiol. 1992;6(3):225-239.

[48] Leins U, Goth G, Hinterberger T, Klinger C, Rumpf N, Strehl U. Neurofeedback for children with ADHD: A comparison of SCP and theta/Beta protocols. Appl Psychophysiol Biofeedback. 2007;32(2):73-88.

[49] Strehl U, Aggensteiner P, Wachtlin D, Brandeis D, Albrecht B, Arana M, et al. Neurofeedback of slow cortical potentials in children with attention-deficit/hyperactivity disorder: A multicenter randomized trial controlling for unspecific effects. Front Hum Neurosci. 2017;11(March):1-15.

[50] Wiedemann M. Infra low frequency (ILF-) neurofeedback. In: Haus K-M, editor. Praxisbuch Biofeedback Und Neurofeedback. 2. Edition. Berlin Heidelberg: Springer; 2015. p. 91-115.

[51] Grin-Yatsenko VA, Ponomarev VA, Kara O, Wandernoth B, Gregory M, Ilyukhina VA, et al. Effect of Infra-Low Frequency Neurofeedback on Infra-Slow EEG Fluctuations. In: Biofeedback. 2018.

[52] Grin-Yatsenko VA, Kara O, Evdokimov SA, Gregory M, Othmer S, Kropotov JD. Infra-low frequency neurofeedback modulates infra-slow oscillations of brain potentials: A controlled study. J Biomed Eng Res. 2020;4:1-11.

[53] Dobrushina OR, Vlasova RM, Rumshiskaya AD, Litvinova LD, Mershina EA, Sinitsyn VE, et al. Modulation of intrinsic brain connectivity by implicit electroencephalographic neurofeedback. Front Hum Neurosci. 2020;14(June):1-13.

[54] Sasu R, Othmer S. Neurofeedback in application to the ADHD spectrum. Restoring the Brain. 2015;(July 2015):231-260.

[55] Othmer SF. Protocol Guide for Neurofeedback Clinicians. 5th. EEG Info Inc.; 2015.

[56] Lubar JF, Shouse MN. EEG and behavioral changes in a hyperkinetic child concurrent with training of the sensorimotor rhythm (SMR): A preliminary report. Biofeedback Self Regul. 1976;3:293-306.

[57] Monastra VJ, Lynn S, Linden M, Lubar JF, Gruzelier J, LaVaque TJ. Electroencephalographic biofeedback in the treatment of attention-deficit/hyperactivity disorder. Appl Psychophysiol Biofeedback. 2005;30(2):95-114.

[58] Logemann HNA, Lansbergen MM, Van Os TWDP, Böcker KBE, Kenemans JL. The effectiveness of EEG-feedback on attention, impulsivity and EEG: A sham feedback controlled study. Neurosci Lett. 2010;479(1):49-53.

[59] Vernon D, Frick A, Gruzelier J. Neurofeedback as a treatment for ADHD: A methodological review with implications for future research. J Neurother. 2004;8(2):53-82.

[60] Vernon D, Egner T, Cooper N, Compton T, Neilands C, Sheri A, et al. The effect of training distinct neurofeedback protocols on aspects of cognitive performance. Int J Psychophysiol. 2003;47:75-85.

[61] Heinrich H, Gevensleben H, Freisleder FJ, Moll GH, Rothenberger A. Training of slow cortical potentials in attention-deficit/hyperactivity disorder: Evidence for positive behavioral and neurophysiological effects. Biol Psychiatry. 2004;55(7):772-775.

[62] Rossiter TR, La Vaque TJ. A comparison of EEG biofeedback and psychostimulants in treating attention deficit / hyperactivity disorders. J Neurother Investig Neuromodulation, Neurofeedback Appl Neurosci Publ. 1995;1(1):48-59.

[63] Lubar JF, Swartwood MO, Swartwood JN, O'Donnell PH. Evaluation of the effectiveness of EEG neurofeedback training for ADHD in a clinical setting as measured by changes in T.O.V.a. scores, behavioral ratings, and WISC-R performance. Biofeedback Self Regul. 1995;20(1):83-99.

[64] Arns M, Ridder S de, Strehl U, Breteler M, Coenen an, Richter JW. Efficacy of neurofeedback treatment in ADHD: The effects on inattention, impulsivity and hyperactivity: A meta-analysis. Clin EEG Neurosci. 2011;30(3):26-33.

[65] Kirk HW. Restoring the brain: Neurofeedback as an integrative approach to health. Restoring the Brain: Neurofeedback as an Integrative Approach to Health. Taylor and Francis; 2015. 1-286 p.

[66] Legarda SB, McMahon D, Othmer SS, Othmer SS. Clinical neurofeedback: Case studies, proposed mechanism, and implications for pediatric neurology practice. J Child Neurol. 2011;26(8):1045-1051.

Chapter 6

Training the Conductor of the Brainwave Symphony: In Search of a Common Mechanism of Action for All Methods of Neurofeedback

Jen A. Markovics

Abstract

There are several different methods of neurofeedback, most of which presume an operant conditioning model whereby the subject learns to control their brain activity in particular regions of the brain and/or at particular brainwave frequencies based on reinforcement. One method, however, called infra-low frequency [ILF] neurofeedback cannot be explained through this paradigm, yet it has profound effects on brain function. Like a conductor of a symphony, recent evidence demonstrates that the primary ILF (typically between 0.01–0.1 Hz), which correlates with the fluctuation of oxygenated and deoxygenated blood in the brain, regulates all of the classic brainwave bands (i.e. alpha, theta, delta, beta, gamma). The success of ILF neurofeedback suggests that all forms of neurofeedback may work through a similar mechanism that does not fit the operant conditioning paradigm. This chapter focuses on the possible mechanisms of action for ILF neurofeedback, which may be generalized, based on current evidence.

Keywords: EEG biofeedback, neurofeedback, electroencephalography, infra-low frequency [ILF], infra-slow oscillations [ISO]

1. Introduction

Neurofeedback is biofeedback for the brain. It is a method that developed out of research curiosity and demonstrated efficacy as a therapeutic modality for improving brain function, although it has taken more than half a century to gain some level of acceptability into the Western medical establishment. Now that it has gained some respectability as a therapy for symptoms of brain and mental health disorders, it is becoming increasingly important to develop a concise theory of the mechanism of action for how neurofeedback causes its effects. There have been many proposed mechanisms, but they are often very narrowly applicable to the particular method of neurofeedback discussed. Furthermore, some lines of investigation have provided evidence to suggest that most of the proposed, straightforward mechanisms are likely incorrect, which may actually explain some of the inconsistent results that have plagued the research literature. Thus, it is important to consider that there may be one overarching mechanism of action explaining how neurofeedback works, which takes into account and applies to all the various methods of neurofeedback.

In this article, several hypothetical mechanisms of action are presented, which were derived from the various methods of neurofeedback, from which a single hypothesis is proposed that attempts to incorporate all of the common features of the other mechanisms in order to more generally explain how all neurofeedback may work.

2. Neurobiology underlying mechanisms of neurofeedback

Since the field of neurofeedback essentially co-developed with our modern understanding of neurobiology, the neurobiological concepts underlying the mechanism(s) of neurofeedback that are presented here are as putative as the neurofeedback mechanisms, themselves. In many ways, the application of neurofeedback, itself, has helped to elucidate the underlying neurobiology. Therefore, it is important not to over-commit to any particular theory or hypothesis, since, in the future, after more information is revealed through rigorous scientific investigation, it may be proven wrong. Scientists must always be willing to pivot from one model to another and not hold too tightly to any piece of "knowledge". Herein, a non-exhaustive description of the current neurobiological foundations in which neurofeedback works to produce its effects are presented.

2.1 The brain as a prediction device

To understand how the brain works, it's necessary to consider some basic functions of the brain and the obstacles it needs to overcome in order to perform such functions. As we all know, the brain is how we perceive and function in our physical reality/world. Everything we are and do is controlled by the brain, such as sensory perception, motor activity (voluntary and involuntary), and cognition/executive functions. Those are the three basic forms of brain function, covering *everything* that we perceive and do. For example, the only way we know that there is a tree in front of us is because we see it (i.e. visual perception), we smell the bark and leaves (i.e. olfactory perception), and maybe we even touch it (i.e. tactile perception), which are all sensory percepts that are processed in the brain. Does that tree really exist outside of the brain? Well, of course, that's a question for philosophers. It's essentially the same question as: If a tree falls in the forest but nobody sees or hears it, does it really fall? Most of us would say, yes, but we have no way to "prove" it (not to mention that the concept of "proof" is also moot in science, but that's another topic), because that requires some sort of observation or exchange of information, which is lacking in the given scenario.

In order to perceive and function in the world, the brain must be able to take in the information from the environment via our senses, process it ("bottom-up processing"), make a decision on how to react, then implement that action (often involving motor output or "top-down processing") [1]. These processes are not instantaneous, but take time, which is a critical obstacle if a quick response is necessary, such as when a lion attacks or a child runs into oncoming traffic. Thus, the brain needs to overcome this relatively slow processing speed by not *re*-acting to the world, but *pro*-acting to it, or predicting it [1, 2]. Prediction is based on prior information about the regularities or patterns in the environment, and it is a key feature of brain function that helps create an accurate representation or model of the sensory environment and any actions required to navigate it [1].

One way in which the brain is able to make predictions is through constant, ongoing activities, such as waves of electrical potentials that are created by oscillations of local field potentials (LFPs) throughout neural networks in the cerebral cortex and subcortical structures, which commonly called brainwaves [3–5].

Anything cyclical or periodic is inherently predictive. While oscillatory activity lends itself easily as a mechanism for temporal prediction (i.e. predicting 'when'), it is also an effective mechanism for predictive coding (i.e. predicting 'what') [3, 6].

Another way in which the brain predicts future activity is through prepotent models of the sensory environment and prepotent models of actions that it forms based on patterns from past situations/scenarios/contexts from which it learned [2]. For instance, in the GO/NOGO task employed for the analysis of event-related potentials, the subject is instructed to press a button every time they see a target for the GO condition, for which the brain creates a prepotent model of the action of pressing a finger down on a button as soon as the target stimulus is presented [7]. Then, once the target is shown, that activity encoded in the prepotent model is easily and quickly performed without much effort since it was pre-planned. However, if a non-target is shown, instead, such as in the NOGO condition, the brain needs to actively put the brakes on that pre-planned action, which actually takes more energy than following through on that action [7, 8]. This brain function is called "response inhibition" and is carried out by the prefrontal cortex, which is primarily an inhibitory cortex, and the primary inhibitory neurotransmitter is gamma aminobutyric acid (GABA) [7, 8].

People whose prefrontal cortices are not fully developed, such as children, teenagers, some young adults, and those who have been diagnosed with attention deficit/hyperactivity disorder (ADHD), have difficulties with response inhibition [7]. These difficulties may manifest in impulsive behaviors, compulsive behaviors, obsessive thoughts, inappropriate remarks or behaviors, etc. [7]. A couple of other examples of brain dysfunction that shed light on the ongoing functions of the brain are automatisms and alien hand syndrome [9, 10]. Automatisms are behaviors that sometimes occur during an epileptic seizure where a set of motor behaviors occur in a particular sequence without conscious thought or agency but may appear purposeful, except for the fact that they do not achieve any particular function and the individual is usually in an altered state of consciousness [9]. Most of the time, those of us who have excellent functioning prefrontal cortices are able to inhibit automatic behaviors that do not have a purpose, but sometimes, particularly when inebriated with mind-altering substances, the brain becomes disinhibited, allowing these behaviors to come out [11].

Alien hand syndrome is probably one of the most fascinating phenomena that reveals brain and body function in a unique and strange way. Some people with intractable epilepsy have surgery that severs the connection between the two hemispheres of the brain – these people are often called "split-brain" patients [12]. Some of these "split-brain" patients, or others who have had a stroke or some other injury or insult, have severed connections between the pre-supplementary motor area (pre-SMA), the anterior and medial cingulate cortices (ACC and MCC), and the sensorimotor cortex (SMC) in either hemisphere [10, 13, 14]. This severance releases the hand contralateral to the SMC lesion from conscious, voluntary control, allowing it to behave as if it has a mind of its own [10, 13, 14]. What is particularly interesting is that this usually does not mean that the hand does nothing and just sits listlessly at the side of the person. Instead, the hand literally behaves as if it has a mind of its own, revealing what a hand would do if it was not told to stop – that is, to grab or clutch [13, 14]. A hand without conscious control will grab anything it "sees" (although it does not actually "see", but since the feedback connection between the visual cortex and the SMC is not severed, information from the visual cortex can directly drive hand motor action since it cannot be inhibited by the dominant prefrontal cortex), such as a glass, a pencil, or even a woman's breast (if you are someone who likes women's breasts) [14]. One of the only ways to get the alien hand to stop grabbing things is to put something in it to hold so it is occupied and unable to grasp anything else [10].

2.1.1 Feedback as general mechanisms of learning and neuroplasticity

A second major characteristic of the brain is its ability to learn and adapt. The neurobiological mechanism of learning is called neuroplasticity, which means that the brain changes – it's considered malleable like plastic, as opposed to something that cannot change easily, like a rock or metal. Learning requires information to be remembered, but it also requires error correction to make sure that the information retained matches the information taught. It is also the mechanism by which we fine tune our performance, which is just another form of learning. For instance, in order to walk, we must move our legs, but in order to know that we are walking, we must get information from our feet that they have met the ground and maybe also from our eyes to see it touch the ground. If all of these bits of sensory information agree they are then integrated together in a feedback loop with the motor action to confirm that what was pre-planned (the motor activity of taking a step) is what actually occurred. This mechanism of feedback is called the sensorimotor loop and is a form of predictive processing, which is a primary mechanism by which the brain is able to respond to the environment and self-correct when errors or perturbations occur [2–5].

Based on feedback loops, predictive processing can be considered a strategy of control systems [2–5]. If the body is a collection of bodily systems, the brain is the control system, defined as a stable system in which its elements interact to preserve stability for both internal control and response to perturbations from external sources [4, 5]. Control systems are characterized by feedback loops, which can either be closed or open [4]. Negative feedback loops in closed systems create oscillatory activity, which are generated in the brain by coupling excitatory and inhibitory neuronal activity in circuits [4, 5]. These oscillatory activities, which can be measured in LFPs, regulate the excitability of the cortex, which regulates the ease with which long-term potentiation (LTP) or learning can occur [15].

A key aspect of learning (i.e. neuroplasticity) is timing. Learning, which requires memory formation, occurs through LTP and is primarily established through spike-timing dependent plasticity (STDP), the conventional form of which is through Hebbian plasticity [16]. Nearly all of our synaptic connections are weakly formed in the first two years of life [17]. After this period of neurogenesis and synaptogenesis, our brains go through nearly two decades of experience-dependent synaptic strengthening and both experience- and neglect-dependent synaptic pruning [17, 18].

According to Hebb's postulate, the strengthening of a synapse requires the precise timing of the activation of two neighboring synapses on the same post-synaptic neuron such that a pre-synaptic signal from the weak synapse is quickly followed by a stronger, post-synaptic signal (coming from an established synapse upstream of it), causing the weaker synapse to appear to co-fire with the stronger synapse, linking them to create an action potential that propagates down the neuronal axon [19, 20]. The timing of these coordinated signals must be very precise, such that the signal from the pre-synaptic neuron into the weak synapse must fire within milliseconds (ms) (generally around 20–40 ms) before the stronger, established synapse on the post-synaptic neuron fires in order for LTP to occur [21]. If the post-synaptic neuron fires first, however, long-term depression (LTD) can occur, which further weakens the synapse, ultimately resulting in synaptic pruning [21]. The typical timeframe for LTD requires that a spike from the presynaptic neuron reaches the weak synapse within 20–40 ms after the spike from the postsynaptic neuron [21]. To complicate the matters, different neuronal populations in different brain regions have their own specific temporal patterns of STDP [16]. The brainwave most frequently implicated in LTP and memory formation is the theta (θ) band [15], which has a phase-amplitude cross-frequency coupling with the gamma (γ) [22] band in the hippocampus. This θ-γ coupling is believed to play complementary functions in memory formation: θ oscillations are involved in encoding whereas γ oscillations (which form ripples) are involved in consolidation [23].

2.1.2 Functional networks

Due to technological advances in imaging, neuroscience has grown exponentially in the past few decades. Using a technique called functional magnetic resonance imaging (fMRI), researchers identified networks of metabolic activity in the brain that work together at the same time (i.e. synchronously) over spatially distant regions, which are connected via white matter tracts [24–27]. These networks are called *functional networks*. Furthermore, there are both task-positive and task-negative networks (a.k.a. resting state networks), meaning that some networks are associated with some sort of voluntary endeavor or task whereas other networks are not associated with a particular activity but are active during times of "rest" or non-directed thought [27]. Eventually, researchers discovered that many of these resting state networks are shared with task-positive networks, with one exception, the default mode network (DMN), which will be discussed in more detail later [25]. Researchers were surprised to see that functional networks continue to be active during times of rest, but those of us who have thoughts streaming through our heads nonstop already knew this about our brains! Even those who do not have thoughts constantly streaming through their brains, however, also have resting state network activities. In fact, the only time when the brain does not seem to be flowing through different functional networks is during states of unconsciousness, or at least the networks during unconsciousness show less connectivity, and the dynamics between networks are slower [28]. The brain continues to switch between resting state networks even during sleep, which is an altered state of consciousness, although their dynamics are also slower than during wakefulness, but not as slow as during unconsciousness or coma [29].

The key aspect of fMR imaging, which differentiates it from regular MR imaging, is the additional signal analyzed, which is the blood oxygen level dependent (BOLD) signal that causes the magnetic resonance to shift in intensity by approximately 1% depending on oxygen-rich or oxygen-poor blood in the region [24, 25]. Essentially, the premise of fMRI that gives it its functionality is the notion that where there is oxygen-rich blood in the brain neural activity is occurring. Furthermore, the BOLD signal fluctuates or oscillates at a typical frequency which is between 0.01–0.1 Hz, or one cycle per every ten seconds to one cycle every 100 seconds [24, 26]. This frequency is the same as the primary infra-low frequency (ILF) that can be measured by electroencephalography (EEG) [26, 30, 31].

Each publication on resting state functional networks seems to characterize a different number and general description of networks, although some networks appear to be consistent across reports, such as the dorsal and ventral attention networks (DAN; VAN), the central executive network (CEN), the salience network (SN), the basal ganglia/limbic network (BGLN), and a series of sensory-related and motor networks [27, 32]. The most consistently characterized network in all reports is the DMN, making it the network in which the brain spends most of its time and energy [30]. The DAN, VAN, CEN and sensory- and motor-related networks are all considered "task-positive" networks as they are associated with specific attentional, executive, sensory, and motor tasks, but they have also been detected at times of rest, as well [26]. The DMN, however, is exclusively associated with times of rest and relaxation, self-reference, and projecting into the mind of others (i.e. "theory of mind"), which are all considered part of the "core self" [30, 33, 34]. The DMN has particular significance in the mechanism of neurofeedback, and its role may be to integrate the self with the three-dimensional body and world in which it inhabits [30, 33, 34].

2.1.2.1 Inter-network dynamics

These functional networks break up into two basic systems of internally-focused (i.e. the DMN) and externally- or task-focused (the so-called "task-positive networks, which essentially refer to all of the other networks) [26]. Since these networks are very dynamic, even at rest, their very characterizations have been relatively elusive, depending heavily on statistical analyses of correlated activities at different nodes or hubs [24]. The characterization of the DMN, however, seems to have great consensus among researchers, revealing it as probably the most important network, which anti-correlates with all of the other networks with little exception [26, 35]. This anti-correlation means that when the DMN is activated, the task-positive networks are deactivated or inhibited [26, 30, 35].

Studies on network inter-dynamics suggest preferential directionalities in these dynamics where certain networks tend to be activated before or after other networks and how different networks modulate the activity of other networks [35–38]. Specifically, the DMN and the SN regulate switching between internally-generated, self-referential/self-focused processing (in the DMN) and externally-generated information processing (such as from the senses) or other cognitive functions that are not self-focused or self-referential (such as math, reading, etc.) in the attention networks, the sensory networks, and the executive networks, etc. (i.e. the "task-positive" networks) [38]. Some of the same network hubs that overlap between the DMN and the SN are also part of the executive networks, which make their differentiation somewhat ambiguous, but both models - where the DMN interacts exclusively with the SN and where the DMN interacts with both the SN and the executive networks – reflect the same underlying mechanism whereby the SN regulates switching between the DMN and task-positive networks, including executive functions [37]. Furthermore, both models make intuitive sense when considering that when you are internally focused you cannot also be externally focused since these are mutually exclusive states.

Network dynamics may be a good neurophysiological measure of neuroflexibility, which can manifest as cognitive and behavioral flexibility, particularly for networks involving the frontal lobes [39]. The level of dynamic switching between networks and their interactions also appears to be more strongly correlated to conscious processes, as opposed to intra-network connectivity, alone [28]. Greater dynamics in functional connectivity correlates with better behavioral responses and results in cognitive tasks, as well as better mental health [40–43]. These results imply that a more flexible brain, one which easily engages and disengages in brain states, is a better functioning brain.

The variation in network participation at inter-network hubs correlates with retrospective self-generated thoughts, which are considered correlates of unhappiness and are precursors for a negative mood [44]. This variation in membership at inter-network hubs, as well as the stability of densely interconnected nodes (considered to be the 'rich club') diminish with age [44–46]. This reduction in network modularity with age suggests less distinct functional divisions between networks, resulting in less information sharing and processing across networks [46]. Furthermore, there are specific changes in inter-network dynamics which also change with age, but the developmental trajectories are specific to the particular interacting networks and may also be specific to certain functions that depend on the particular activities and interests of the individual over their lifetime [45, 47].

2.1.3 Glial cells

Neurons get all of the attention when it comes to the brain and the nervous system, but they can only do what they do because glial cells provide protection, nutrients, neurotransmitters, insulate axons (creating myelin), assist in synapse formation and remodeling, protect against foreign attack, clean up extracellular debris, maintain structural integrity of the tissue, etc. [48–50]. The ratio of glial cells to neurons in the human brain has typically been reported as anywhere from 4 to even 50, although these numbers are inaccurate due to the region-specific ratios, while total numbers of neurons and glia have a ratio of nearly 1:1 [51]. Despite the variability in these regional ratios, an argument could be made that glial cells, as opposed to neurons, are the most important cells of the brain.

The term, *glia*, is derived from the Greek word meaning glue, and reflects the original function that these cells were believed to do, which was essentially holding neurons together in the brain like glue [48, 49]. In recent years, however, scientists have discovered that these cells provide substantially more functions than just structural integrity of the brain. For instance, astrocytes or astroglia, which are named for their star shape, help create and maintain the blood brain barrier, regulate the formation, maturation, maintenance, and stability of synapses, and regulate specific neuronal network activities through the inhibition of local, non-specific activities [18, 52, 53].

Astrocytes, in fact, play a central role in regulating neuronal activity through metabolic coupling and neurotransmitter recycling [52]. Through their foot processes that wrap around the endothelium of the capillaries and their intimate contact with synapses (creating the *tripartite synapse*), astrocytes play crucial roles in neurovascular coupling [54]. This coupling allows astrocytes to regulate the metabolic activity of the neurons associated with these synapses through the release of ATP, as well as the removal of metabolites and the recycling of ions and neurotransmitters into the synapse [52, 53].

Recent studies on astrocytes have revealed their heterogeneity in the human brain, which may be as diverse as all of the different neural circuits and networks [55, 56]. In fact, astrocytes are implicated in the development, plasticity, and function of neural circuits [18]. Furthermore, astrocytes have bioelectrical properties that are created by calcium fluctuations across its membrane, which couple with neuronal firing and are likely the source of LFPs, which create the brainwaves that can be detected by EEG [57, 58].

The two other glial cell types in the central nervous system, microglia and oligodendrocytes, play important roles in neurodevelopment, neuronal signaling, neuroplasticity, and neuroprotection [49]. Oligodendrocytes create the myelin sheath on neurons, allowing for faster and more efficient propagation of the action potential down the axon, while microglia are critical for neuroprotection as the resident phagocytic immune cells of the brain [49]. Microglia also play critical roles in neurodevelopment and neuroplasticity (particularly during synaptic pruning), and their dysfunction is implicated in the etiology of many neurodevelopmental disorders and neuroinflammation [59]. Despite the critical functions that these cells play in the brain, no significant and/or unique roles have been ascertained for them at this time in the possible mechanisms of neurofeedback.

2.2 Clinical benefits of neurofeedback training

There have been many reports on the clinical benefits of neurofeedback training but describing in detail all of the studies on the efficacy of different neurofeedback methods in improving the multitude of brain-related symptoms reported in the literature is beyond the scope of this chapter. Therefore, the reader is referred to some recent reviews of the subject [60–64]. Briefly, a general description of neurofeedback effects that modify different symptoms are presented here.

2.2.1 Seizures

The first clinical effect of neurofeedback was the reduction of seizure incidence, duration, and severity in cats, which was then recapitulated in humans [65–67]. This original protocol, which trains the so-called sensorimotor rhythm (SMR), which was originally developed for cats, is still used for the treatment of seizures in humans today, but the seizure-reduction benefits are not exclusive to that method and can be achieved using other forms of neurofeedback, as well [63, 68, 69]. Furthermore, the method is generally the same no matter what kind of seizures the subject has, including psychogenic non-epileptic seizures (PNES), unless it's a focal seizure, which could require specific electrode placements to target the focus [68, 69, personal experience].

2.2.2 Attention/focus

Most of the research in the field of neurofeedback has centered around its benefits for improving executive functions in people who have been diagnosed with ADHD [60, 65, 66]. In fact, in 2013, the American Association of Pediatrics endorsed neurofeedback at level 1 effectiveness for the treatment of ADHD, which is the same level endorsed for ADHD medications [70]. Due to the fact that the literature is over-saturated with these reports on the efficacy of neurofeedback for ADHD, the reader is directed to some excellent, recent reviews [71–73]. The main point is that neurofeedback has been used to improve symptoms of ADHD, particularly executive dysfunctions such as inattention and difficulty shifting tasks, etc., but also hyperactivity and impulsivity [71–73]. Improving symptoms of ADHD, however, is merely the tip of the iceberg of what neurofeedback training can achieve.

2.2.3 The multitudes of symptoms improved by neurofeedback

It's truly difficult to describe all of the symptoms, behaviors, and other effects that neurofeedback has affected, and even more difficult to show evidence of these effects via randomized, controlled trials (RCTs), which are the gold standard for determining evidence-based practice in our Western healthcare system. However, despite the challenges of designing, obtaining funding, implementing, then publishing such studies, the field of neurofeedback has continued to progress due to the very real benefits that clients continue to gain from it, spreading the word to others who then try it and also witness its benefit [74, 75]. Of course, the issue is that this method of utility and expansion of the field is unsustainable in a healthcare system dependent on insurance where insurance will only pay for what the Western medical establishment considers evidence-based practice.

Clearly, evidence-based practices are ideal in order to demonstrate efficacy and build trust in the field, but the only way to get evidence is to practice. It's been a bit of a Catch-22 to reach the level of evidence-based practice when grant lending agencies have been reluctant to fund studies in the field of neurofeedback, but in order to get funding, the clinicians need to establish some level of efficacy of the practice to recruit enough subjects for the gold standard RCTs. In the meantime, the clinicians have done their best on their own to optimize neurofeedback for its benefits to their clients in the absence of funding for these gold standard studies [66, 74, 75].

In addition to its benefits in reducing seizures and improving ADHD symptoms, some of the multitude of benefits of neurofeedback training are reductions in headaches, migraines, anxiety, irritability, post-traumatic stress symptoms, etc., and improvements in sleep regulation, pain management, mood, peak performance, etc. Most of these effects have been documented in case studies and/or experimental trials that are less rigorous than RCTs [61–63, 74, 76]. Essentially, any function of the brain can be modulated using neurofeedback, which theoretically means that everything we do, think, and feel can be improved using neurofeedback.

3. A very brief history of neurofeedback

Neurofeedback originated as EEG biofeedback and developed nearly at the same time as EEG, itself, beginning in the 1930s with Hans Berger and his colleagues toying around with this new machine to watch how their brainwaves changed with different perturbations [60, 65, 66, 77, 78]. However, neurofeedback as a therapy did not start until the 1960s after Barry Sterman discovered that neurofeedback training could protect cats against jet fuel-induced seizures [67]. Around the same time, another scientist, Joe Kamiya, was also conducting neurofeedback using alpha (α) training, which he found was able to reduce anxiety [77, 78]. After publishing an article about his experience with α training, public awareness and interest was piqued, so more people decided to try it out. Some people who tried α training had a spiritual experience, which is great for them, but, unfortunately, it did not bode well for the reputation of the emerging field of neurofeedback since the spiritual effects were seen as contrary to true medical treatment or therapy [66, 77, 78].

Although the remainder of the history of neurofeedback is quite interesting and provides a good background for why the field is so divergent today (primarily due to in-fighting between pioneering researchers), it is not within the scope of this article to present. Therefore, the reader is directed to many other resources where they might find a history of the field [65, 66, 77, 78]. The main points here are that neurofeedback began as EEG biofeedback, which developed out of pure curiosity about how one might be able to control their brainwaves through visual feedback of their activities, and the original methods that were developed were α training and SMR/beta (β) training.

3.1 A very brief history of electroencephalography

Neurons communicate quickly through electrical impulses that travel down long axon tails, ending at a synapse, which is where the electrical signal changes to a chemical signal between neurons [19]. Neurons can also communicate directly through electrical synapses, as well, although this is not as versatile of a signal (i.e. it cannot be regulated to the same degree as a chemical signal) [19]. We can detect the electrical signal in the brain using EEG, which is a technique that detects changes in electrical potential via electrodes that are placed on the scalp [7]. Since the electrodes are several millimeters away from the cerebral cortex, and their surface area is thousands-to-millions of times larger than the surface area of a single neuron, they are actually picking up the summed electrical field potentials from millions of neurons and their surrounding glial cells, both cell types of which can establish an electrical potential across their membrane [7, 52, 57].

Hans Berger was the first person to publish an article with the first human EEG in 1929 [65, 77, 79]. After placing an electrode on the back of the head, Berger observed a wave of electrical potential with an approximate frequency of ten cycles per second or 10 Hertz (Hz), which he coined as α waves [60, 77, 79]. After this initial discovery, several other brainwave bands were discovered, which will be discussed in more detail in the next section [77].

3.1.1 Brainwaves

As mentioned in the previous section, the invention of EEG led to the discovery of ongoing electrical oscillatory activity in the brain, which could be detected using electrodes placed on the scalp. Following the discovery of α waves, β waves (≥ 13 Hz) were then discovered, then delta (δ) waves (1–3 Hz), and θ waves (4–7 Hz) [77]. γ waves are a subset of very high β waves (> 30 Hz) [79]. Collectively, we have dubbed these *brainwaves* [57]. These brainwave bands show specific patterns of activity at specific times, locations, and during specific brain activities, although their precise functions are not abundantly clear [57, 79]. Furthermore, the origins or oscillators that generate the brainwaves are also not very clear or easily defined, although studies suggest key oscillatory roles for the thalamus, the reticular activating system (RAS) of the brainstem, and specific layers of pyramidal neurons and astroglia in the cortex [57, 79, 80].

Characterizations of the classic brainwave bands are primarily derived from sleep studies and studies of patients with epilepsy [81]. Early neurofeedback studies also attempted to functionally characterize these brainwave bands, but their functional characterization has been elusive due to the promiscuity of associated activities [57, 79, 81]. One intriguing theory posits that the different brainwave bands are evolutionarily related and similar in categorization as the *triune* model of brain structure and function [82]. In this model, δ oscillations are analogous to the brainstem's basic functions, representing the evolutionarily oldest structures and functions of the brain, while θ and α oscillations are analogous to the limbic system and basal ganglia, dominating in lower mammals, and the fastest brainwaves, β and γ, are analogous to the neocortex, which is the evolutionarily newest structure with associated functions such as higher order cognitive processing and self-awareness [82]. Although this model is very intriguing, it remains to be substantiated with experimental evidence.

A more complex picture of brainwave activities has emerged in recent years that describes different ways in which brainwaves interact with each other through cross-coupling [22]. Much like functional networks, researchers discovered different coupling patterns in ongoing activity that spatially organize in networks similar to and often overlapping with the intrinsic functional networks that were discovered by fMRI [22]. These intrinsic coupling modes (ICMs), as they have been dubbed, demonstrate more precise correlations with function than single brainwave bands, indicating a much more richly complex structural and functional architecture to the electrophysiology of the brain [22]. For example, a specific difference in visual perception – whether two lines appear to bounce away from each other or pass through each other – correlated with the percentage of coherence in a β phase ICM in specific visual processing regions (i.e. those with higher β coherence perceived bouncing, whereas those with lower coherence perceived passing) in a study using the bounce-pass paradigm [83]. Delineating function to specific brainwave bands is challenging because: (1) the function can be highly specific, which means each specific brain function would require definition by experimentation, (2) each brainwave band is associated with many different specific functions with little generalizability between them, and (3) functions of brainwaves are spatially and temporally specific, and may be specific to certain cross-frequency couplings [22, 57, 79].

3.2 Modern neurofeedback

Today, there are many methods of neurofeedback using different technologies, data, and protocols (for some reviews, see [61, 63–65, 84]). Different researchers group the methods in different ways. For instance, some researchers group the methods according to their neuroimaging technology (i.e. EEG or fMRI) [75]. Other researchers group the methods according to the temporal structure of the data, such as methods that use *discrete* events (e.g. frequency training, fMRI, etc.), in which there are periods of target activity and periods of rest, and methods that use *continuous* "events" (e.g. ILF), in which the target signal is continuous and the goal is not necessarily to modulate its activity [65]. Two general categories of methods have emerged in the field of neurofeedback, those that employ a *directive*, operant conditioning approach, which require explicit awareness and learning by the subject, and those that are *non-directive* in nature, employing a passive approach, which only require *implicit* learning of self-regulation of which the subject is likely unaware [84, 85]. The directive methods are the most abundant and commonly used neurofeedback methods, and the theories behind them are the easiest to understand and explain. Thus, the directive methods are presented first.

3.2.1 Directive methods

Directive methods of neurofeedback refer to the methods that require the clinician to *direct* the subject to consciously control specific brainwaves by learning what it feels like to modulate the brainwaves during the session. These methods have also been called *explicit* methods of neurofeedback, which refers to the fact that the subject learns to become consciously aware of their brain states [84]. The purpose of the directive methods is to learn to voluntarily regulate specific brain activities and, by reinforcing them, these specific activities increase, altering the pattern of brain activity. Thus, there are two outcomes for directive methods of neurofeedback: (1) how well the subject *learned* to regulate their own specific brain activities, and (2) improved symptoms and behaviors [84, 86].

These directive methods are based on assumptions of how the brain *should* work, using these assumptions to tell the brain what it needs to correct and how to correct it, using an operant conditioning model to "fix" the broken brain. Developed by the pioneering behaviorist, B.F. Skinner, operant conditioning is a method of learning that is based on positive reinforcement and punishment or rewards and inhibits, respectively [87]. When the subject does the desired behavior or learns the desired skill, then they receive some sort of reward, which may or may not have anything to do with the new behavior or skill. Since the reward is desirable, the subject will then repeat such behavior or skill with greater frequency. The behavior or skill is considered learned when the subject can perform it even without the reward. The original neurofeedback methods and most methods still employed today use this operant conditioning model to both direct the procedure as well as to explain how neurofeedback works [61, 64].

3.2.1.1 Conventional EEG-based training

As mentioned previously, the original neurofeedback training protocols used operant conditioning to train the brain to modulate the power of conventional EEG bands, such as the SMR, which is in the low β band, and α training [65]. These protocols continue today, but now they have expanded to include more brainwaves and different properties of the EEG, such as coherence (also called "synchronization") and phase [63]. With the development of quantitative EEG (QEEG) methods, where a cap of 19 electrodes on the head records all channels simultaneous and the activities in each region can be quantitatively compared to others, databases have been created using both normative data from healthy controls as well as comparative data from brains with different symptomologies and diagnoses [7, 74]. These databases can be used as a resource to determine the significant differences between a subject's QEEG activity and that in the normative database, providing a statistical score called a "z-score", which indicates how significantly different the activity is and in what direction (i.e. increased or decreased) [7, 74]. Some methods utilize this resource as part of their protocol to train the brain closer to the norms, presuming that the most common QEEG patterns are preferred for better functioning [60, 61, 63, 88, 89]. These additional methods are discussed briefly in this section, although full and complete descriptions are not within the scope of this article and the reader is directed to the previously mentioned review articles and their references within for a more in-depth understanding.

3.2.1.1.1 Frequency training

As mentioned, there are a plethora of conventional EEG-based neurofeedback protocols. These protocols range from single frequency trainings to whole frequency-band trainings to two or more frequencies or frequency band trainings at the same time, etc., training in different directions –"up" or "down" - meaning increasing or decreasing the power and incidence of that frequency or frequency band [61, 62, 64]. Sometimes the protocols are designed based on differences in the QEEG from the normative database, while other times the protocols are determined based on symptom presentation [7, 60]. Some classic frequency training protocols are SMR for epilepsy, as previously mentioned, but also for ADHD, α-θ training for trauma reorganization, midline θ training and θ/β training for focus/concentration, as well as training each individual conventional EEG band for various conditions [7, 60–64, 74].

3.2.1.1.2 Other EEG-based methods

As mentioned at the top of the section, other EEG-based methods of neurofeedback use different aspects of the EEG data to train the brain. For instance, QEEG-based training uses z-scores as the substrate for the feedback, where the brain is rewarded as the QEEG pattern normalizes closer to a z-score of 0 [60, 61, 63, 88, 89]. The issue with this method is that it presumes that the most common QEEG patterns in healthy individuals reflects the best, most optimal pattern of brain function. However, just because something is typical or common does not mean it is the best. In fact, one might hypothesize that anyone who has an exceptional brain, maybe with a very high I.Q. or great talent, might also show differences from the norms – quite possibly significant differences, in fact – but we might presume that these particular differences confer their exceptional abilities as opposed to pathology.

Another method that uses a full cap of 19 electrode placements is low resolution electromagnetic tomography [LORETA] training [61, 75, 88]. LORETA is a method by which QEEG data is analyzed using blind source separation methods like independent component analysis (ICA) to localize the neural source of the signal in the brain and projects that source onto a three-dimensional map of the brain, including subcortical nuclei [7, 88]. The EEG substrates that can be monitored via LORETA are EEG band power, coherence, and/or phase, as well as z-scores [60, 61, 90].

Low-energy neurofeedback system (LENS) is another method of neurofeedback based on EEG data, but it diverges from all other methods in that it actually delivers a very weak electromagnetic pulse into the subject's brain while they lay motionless with their eyes closed [61]. LENS is a very quick treatment and does not require months of sessions, but it is the only neurofeedback methodology that activity adds an exogenous signal to the subject, so it may be considered less "non-invasive" than the other techniques [60]. The basic hypothesis behind LENS is that it perturbs the brain's typical activity by delivering the weak electromagnetic pulse in order to get it "unstuck" [60].

3.2.1.2 Hemodynamics-based training

Since neuronal activity is tightly coupled to hemodynamics (blood flow), several forms of neurofeedback use hemodynamics as a measure of brain activity in specific regions of the brain as substrates for feedback [61, 63]. The most well-known method is fMRI training, which monitors the BOLD signal in brain structures with high spatial resolution [91–93]. Other hemodynamic-based methods that are less well-known are hemoencephalographic (HEG) training and functional near-infrared spectroscopy (FNIRS) training, which have less spatial resolution than fMRI but greater than that of EEG [67, 94, 95]. FMRI neurofeedback has the greatest number of gold standard RCTs, likely due to the fact that the principal investigators on those studies typically are medical doctors and tend to receive more funding than non-medical doctors [91, 93].

3.2.2 Non-directive methods

Non-directive methods of neurofeedback, which have also been called *implicit* neurofeedback, do not rely on directions from a clinician to the subject, nor do they rely on the subject to consciously regulate their brain activity [84, 85]. In fact, the subject does not need to do anything to receive the benefits of these methods of neurofeedback as they are entirely passive processes. These methods have only been in development over the past 10–15 years or so, partly due to their dependency on technological advances of the neurofeedback equipment and software [65, 96, 97].

3.2.2.1 Infra-low frequency training

ILF training grew out of conventional EEG frequency training [65, 74, 76, 97]. It is based on rewards and inhibits, which are aspects of conventional EEG frequency training, but these concepts no longer make sense at such low frequencies (the current Cygnet software, version 2.0.7.4, can now filter out frequencies as low as 0.0001 mHz, which is approximately one cycle per 116 days). Even though there are biorhythms that are as slow and slower than the current limit of detection by this EEG amplifier [95], there are no known neural or glial origins of these very slow oscillations, causing controversy over the source of the signal [65, 76, 97]. However, a recent study demonstrated that 20 sessions of ILF neurofeedback training increased the power of all of the ILFs ($\leq$ 0.1 Hz), including the typical peak around 0.01–0.1 Hz, which is called the infra-slow oscillation (ISO) and correlates with the BOLD signal [98, 99].

One of the differences between ILF and conventional EEG training is that ILF training cannot work through an operant conditioning model since there are no discrete events to reward [65]. Interestingly, though, there is a complex multi-frequency band algorithm of inhibits that follow the individual subject's regular pattern of EEG activity using thresholds that reset moment-to-moment to allow for approximately 95% success rate (which can be modulated in the software) of the signal remaining below the threshold [100, 101]. The inhibits are a summation of over-threshold signals from the different conventional brainwave bands, causing the screen to gray out and the sound volume to reduce, which essentially tells the brain not to make any sudden moves or EEG spikes [100, 101]. These inhibits, therefore, function to stabilize brain activity while the "reward" or training frequency provides continuous information on cortical excitability.

3.2.2.2 Dynamical neurofeedback (i.e. NeurOptimal®)

The final method of neurofeedback that requires some mentioning due to its proliferation in recent years uses a nonlinear dynamical approach and is implemented with equipment and software called NeurOptimal® [102]. Unfortunately, there are no peer-reviewed publications that describe this method and only a very few publications were found of studies using it [103, 104]. Thus, it is hard to describe this method due to the fact that the details on what it is are murky and based entirely on non-peer-reviewed content on the NeurOptimal® website [102]. Descriptions of the method, however, suggest a similar form of training as described for the inhibits used in ILF neurofeedback in the previous section.

Based on theories developed by Val Brown, who originally came up with a five-phase model of neurofeedback training [105], and empirical experimentation performed by his wife, Susan Cheshire Brown, NeurOptimal® claims to train neural dynamics by providing feedback information on how these dynamics change [102]. This information is obtained through threshold boxes on either filtered or fast Fourier transformed EEG frequency bands that set both a minimal and a maximal power for each (several frequencies or frequency bands are monitored at once), and violations of these threshold boxes drive the feedback [102]. Essentially, when the signal power is outside of the threshold box, it causes the sound to reduce or stop in a manner to inhibit this change in activity [102]. Due to neither training frequencies directly "up" or "down", the creators claim that this method of passive observation of sudden activity changes is 100% safe and has no adverse effects [102]. However, this method requires substantially more research to verify its effects and more transparency in its process to allow other researchers to investigate its potential mechanism of action, as well.

4. Hypotheses for the mechanism of action of neurofeedback

These different methods of neurofeedback have led to the development of different hypotheses for how neurofeedback works to improve brain function. It is possible that different methods of neurofeedback work through different mechanisms to produce their effects, but the simplest model would be that all neurofeedback methods work through the same or a similar mechanism of action. Thus, the true mechanism of action of neurofeedback has yet to be determined and collectively confirmed. Here, several general hypotheses based on different methods of neurofeedback are presented, then the common denominator(s) of these hypotheses is discussed more thoroughly as a possible true underlying mechanism of action of neurofeedback, awaiting experimental designs to test and confirm its validity.

4.1 Hypotheses from directive methods: operant conditioning

Directive methods of neurofeedback, which use operant conditioning, rely on the clinician's assumptions about what is wrong with the subject's brain and how to fix it. This seems like a tall order, but it is the primary paradigm by which the Western medical establishment operates. Medical doctors spend four years of their education memorizing everything there is to know about the body, what goes wrong, and how to fix it. Thus, this same paradigm has been applied to neurofeedback with varying success.

The problem with this "clinician knows all" paradigm, however, is that it does not leave room for the unknown. In general, when a medical doctor cannot find anything "wrong" with a patient using any of their tests, they tell the patient that there's nothing wrong and that whatever is happening to them is psychological or psychosomatic. There are even fancy terms for findings that the doctor cannot explain, such as "idiopathic", which just means that the doctor knows that there's a problem, but no one knows what it really is. Another issue with this paradigm is that knowledge changes and what if what we thought we knew for sure turns out to be wrong or at least sufficiently incomplete? In that scenario, whatever the doctor or clinician says and/or does based on this faulty knowledge will either have little to no benefit or may even be harmful for the patient. Of course, this scenario has happened in the past, as well – for example, there was a time when doctors recommended smoking cigarettes as a treatment for asthma [106], but now we know that smoking cigarettes is not only *not* healthy, but actually harmful, hastening disease and death.

There are basically two types of directive methods, those based on brainwave frequency information and those based on blood flow or BOLD activity. In general, brainwave data has high temporal and low spatial resolution whereas blood flow or BOLD data has high spatial but low temporal resolution [107]. However, blood flow and BOLD data are more directly and generally associated with overall brain activity in a region, while all regions of the brain have some sort of brainwave activity at all times, so the level and type of brain activity visualized in the EEG is differentiated by the particular brainwave frequencies, their amplitudes, and their cross-frequency coupling [22, 79]. Understanding the functional meaning of brainwaves in any region at any point in time has been the focus of much research for decades and remains incomplete and somewhat ambiguous [22, 57, 79, 81, 108]. Thus, neurofeedback that is based on the functional meaning of specific brainwaves identified from specific locations on the scalp remains controversial due to the controversial and non-consensus nature of the underlying science.

4.1.1 Fixing bad brainwave patterns

The first and primary hypothesis of EEG biofeedback is based on the idea that there are normal and healthy patterns of brainwave activities during rest and/or tasks and that mental illness is caused by abnormal brainwave patterns [7]. There is evidence that certain brainwave patterns, either at rest or during a task, are associated with specific symptoms and mental disorders, but their causal roles are far from established [7, 109]. Due to the fact that altering these brainwave patterns has shown moderate success in diminishing such symptoms, these hypotheses have gained some traction in the biofeedback field [7, 63]. However, there are alternative explanations for the success of the neurofeedback that are not consistent with the hypothesis of a causal role for "bad brainwaves" in symptoms of mental illness.

One hypothesis that has driven much of the neurofeedback field for use in improving attention for people with ADHD is that β waves are associated with focus and attention [7]. Although this hypothesis is likely true at times since β waves do correlate with both activating (i.e. glutamate-mediated) and inhibiting (i.e. GABA-mediated) neurotransmission [7], it is not always the case that [beta symbol] activity reflects a focused or attentive brain. β activity has also been associated with anxiety/agitation, so increasing β activity to improve executive functions could backfire if it increases anxiety, which, of course, inhibits executive functioning [7]. From a technical perspective, using β activity to drive feedback may also result in modulation of muscle activity due to electromyographic [EMG] artifacts in the β range of the EEG, which tend to be of greater amplitude than true β brainwave activity [64].

A more nuanced view takes into account θ power, as well, and its ratio with β power in the frontal cortex, called the θ:β ratio (TBR), which is supposedly higher in ADHD brains compared to non-ADHD brains [110]. As the only FDA-approved biomarker for ADHD, the TBR should then be able to differentiate between individuals with ADHD and those without the diagnosis [111]. Unfortunately, there is only a very short window in childhood when the TBR can efficiently be used to distinguish between children with ADHD and those without ADHD, and it cannot differentiate ADHD from non-ADHD adults at all [110–112].

The unreliability of the TBR as a biomarker for ADHD symptoms and behaviors is exemplified in the fact that, although θ-β neurofeedback training can result in improved ADHD symptoms and behaviors, individual learning curves and increases in β power do not correlate with these behavioral outcomes, meaning that, although the neurofeedback training did work to improve ADHD symptoms and behaviors, it was not because the subjects normalized their TBR [113]. Other studies support this notion that frequency training neurofeedback *learning,* which is the measurement of how well subjects are able to consciously control their brainwaves, does not always correlate with behavioral improvements [86, 114]. These results indicate that this hypothesis of "fixing brain wave patterns" is incorrect and that there is a better explanation for the effects of neurofeedback training.

4.1.2 Increasing brain activity in specific regions

Another very simple hypothesis for the mechanism of action of neurofeedback training is the idea of increasing brain activity in specific brain regions. This hypothesis is most applicable to the hemodynamic training methods like HEG, FNIRS, and primarily fMRI neurofeedback, which has the greatest spatial resolution [107]. Specifically, the idea is to provide information on where the brain is active in real-time, and to reward the subject when the brain is active (or inactive) in the target region, such as the prefrontal cortex or the amygdala, training up for attention and down for emotional calming, respectively [93]. The straight-forward hypothesis here is that the different brain regions are associated with different functions and activating or inactivating these regions should increase or decrease those functions, respectively. There has been no evidence to negate this hypothesis, but the technology used for the greatest spatial resolution (i.e. fMRI) is not practical for use in a therapeutic setting.

4.2 Hypotheses from non-directive methods: State-based shifts

Hypotheses for the mechanism of action of non-directive methods of neurofeedback have been more challenging to define and test compared to the directive methods due to the fact that they do not follow a straightforward operant conditioning model. These hypotheses are typically fairly general, involving the concept of calming the body system by shifting from a sympathetic nervous system (SNS)-dominant state to a parasympathetic nervous system (PNS)-dominant state, improving self-regulation of regulatory biorhythms, such as the circadian and ultradian rhythms, and increasing network system dynamics and stability.

4.2.1 Biorhythm regulation

In addition to brainwaves, there are a plethora of biorhythms in the body, not to mention the global rhythms of nature, itself. These rhythms span several orders of magnitude from sub-daily rhythms (ultradian) to daily (circadian), monthly (menstrual), seasonally (circannual), into yearly, decadal (10 years), and so on [95]. If these rhythms were approximated as frequencies, these longer biorhythms fit neatly into the ILF range at 0.1 mHz (ultradian), 0.01 mHz (circadian), 0.0025 mHz (menstrual) and 0.0001 mHz (circannual, approximately 3 months), which are all within the parameters of ILF neurofeedback training using the current version of Cygnet® (v.2.0.7.4, beemedic.com, 2021). Therefore, one of the hypotheses for the mechanism of ILF neurofeedback, specifically, is that it trains and improves these biorhythms using intrinsic error correcting through feedback.

When the software could only reach as low as 0.1 mHz, which translates to an ultradian rhythm, David Kaiser proposed a mechanism through which ILF neurofeedback trains the ultradian rhythm, which is created by astrocytes [48]. Although astrocytes may be a key contributor to the mechanism, it is unlikely that the ultradian rhythm is the primary mechanism through which ILF neurofeedback works, since, as the software continues to improve, most subjects tend to have optimal training frequencies at the bottom of the register, which changes with nearly every update of the software (although it is unlikely to continue to change indefinitely) [115]. Furthermore, an electrophysiological signal corresponding to these slower, lower frequencies (< 0.1 mHz) has yet to be described, particularly with a neural or glial source. These slower biorhythms are created through clock gene feedback loops which regulate cascades of signaling pathways throughout the body, including hormonal regulation and even telomere length throughout life [95, 116]. Thus, it is unlikely that direct regulation of long, slow biorhythms, such as the ultradian, circadian, and circannual rhythms, are responsible for the effects of neurofeedback.

4.2.2 Polyvagal theory and the default space model of consciousness

Polyvagal theory is a theory describing the tripartite development of the autonomic nervous system and the functions of the resulting subsystems [116]. Through three phylogenetic stages in evolution, three subsystems of the autonomic nervous system have arisen in higher-order organisms: the sympathetic nervous system, which produces the "fight-or-flight" response, and two branches of the parasympathetic system via the vagus nerve, one branch corresponding to the ventral vagus nerve, which produces a social communication system through facial expression, vocalization, and listening, and the other branch corresponding to the dorsal vagus nerve, which produces the "freeze" response if attempts to fight or escape do not resolve the threat [116–118]. The theory further postulates that many neuropsychiatric disorders may be due to low vagal tone in one or more of these branches, particularly the ventral vagal neurons in the nucleus ambiguus [118, 119].

A key feature of the polyvagal theory is that it integrates neural circuits and rhythms of the brain with those of the heart and gut, which are relevant to all biofeedback modalities [119]. Another, similar theory that integrates these visceral functions with brainwave activities is a theory of consciousness called the Default Space Model, which proposes that at very slow oscillations, the brain synchronizes with the cardiorespiratory rhythm, activating the DMN, which integrates external and internal sensory input to create a three-dimensional conscious experience [30, 33]. Thus, according to this hypothesis, ILF neurofeedback induces the brain into the DMN, and engages the ventral vagal system, which promotes calming, self-soothing, and socially engaging behaviors [34, 118].

4.2.3 Control system dynamics and stability

Despite the lack of information on the NeurOptimal® method of neurofeedback, its description as a nonlinear dynamical neurofeedback system has been helpful in elucidating how neurofeedback training may interact and influence the dynamical system processes of the brain. As mentioned in Section 2.1.1 on feedback mechanisms in learning, control systems, such as the brain, are characterized by feedback loops, and feedback loops create oscillatory activities, which are inherently dynamic [4]. By exercising these dynamics through neurofeedback, the brain resonates with itself, causing an amplification of this activity and creating both greater stability as well as increasing dynamics [4, 5].

4.3 The global hypothesis: self-resonance and system dynamics with microstate stability

There are two aspects that are common to all of these hypotheses: (1) self-resonance from the self-referential feedback, itself, and (2) increased system dynamics with microstate stability. All biofeedback provides self-resonance, which is subjectively calming and comfortable for the subjects [4, 120]. No matter the modality of biofeedback, the resonance of brainwaves or heart rate or baroreflex fluctuations synchronize with each other, ultimately settling comfortably in a PNS-dominant or ventral vagal state, which is restorative, clarifying, and energy-efficient [30, 33, 116, 120].

The second part of the global hypothesis refers to the increasing dynamics of the neuroelectric functional network system. Each method of neurofeedback relies on sensory information to provide the feedback, which requires the brain to enter into the various sensory functional networks in order to process it. However, the information, itself, may not reflect activity in the sensory networks, and the calming aspect of the self-resonance will activate the DMN, which is mutually exclusive with the sensory networks [26, 30, 35]. In fact, several studies show that neurofeedback training increases DMN connectivity, supporting this hypothesis of activating the DMN [34, 121–123].

Acquiring the neurofeedback, itself, requires dynamic shifts between task-positive networks and the DMN, thus strengthening this shifting ability or network dynamics. These effects can be seen in emergent subnetworks that are present immediately after neurofeedback training that combine hubs from the SN, basal ganglia/reward network, and the visual network (presumably due to visual feedback) [85]. Furthermore, the specific brain location or brainwave that is the substrate for the feedback strengthens and stabilizes that brain activity or microstate [5, 123]. They are called 'microstates' because they are short-lived due to the nature of dynamics, but their stability is in the strength of their connections (in the case of functional networks) or peak power intensity (in the case of brainwaves), conferring the brain resiliency against perturbations [5]. These effects translate to improved brain function in the same manner that increased inter-network dynamics improves brain function, as described in Section 2.1.2.1.

Essentially, this common hypothesis combines general mechanisms of biofeedback that confer a calm, parasympathetic-dominant state with specific mechanisms of neurofeedback that exercise inter-network dynamics while stabilizing intra-network connections. As described earlier, these increased dynamics result in improved cognition and mental wellbeing, while the increased stability results in greater resilience. Thus, all methods of neurofeedback improve overall brain self-regulation, where some methods may achieve this more globally and other methods achieve it through more specific detailed aims, such as training very specific brainwave patterns or regions of activity.

4.3.1 The regulatory functions of the infra-slow oscillations

As mentioned in Section 3.1.1, different brainwaves can interact with each other and become coupled, meaning that their activities correlate [22]. These correlations may occur according to phase (where the phase of the slower brainwave regulates the discrete activity of the faster brainwave) or envelope (where the envelope of the slower brainwave modulates the amplitude of the faster brainwaves) [22].

Thus, these forms of cross-frequency coupling of electrical oscillations in the brain suggest that information about one brainwave automatically provides information about another, usually slower, brainwave, which is embedded in its fluctuating activity. However, the resolution of the information of the slower brainwave embedded in the information of the faster brainwave is lower than if the slower brainwave was observed directly.

Although there are many reports on the significance of the cross-frequency coupling of conventional EEG brainwave bands, such as θ-γ in the hippocampus or δ-θ-γ in the auditory cortex or δ-α in the left and right homologous regions of the attention networks, these are short-lived interactions that are both spatially and functionally-specific [22]. One cross-frequency interaction that is constant, however, is between the ISO (typically between 0.01–0.1 Hz) and all of the faster, conventional EEG brainwaves, including δ through γ bands (~1–40 Hz) [99]. This interaction is a phase-amplitude coupling where the amplitudes of all of the faster frequencies are regulated by the phase of the ISO [99].

Studies show that the ISO and the BOLD signal from fMRI correlate and may be part of the same activity, representing the fluctuations of oxygenated and deoxygenated blood [30, 31]. Since neither oxygen nor blood, themselves, create LFPs, the source of the ISO is likely calcium fluctuations across astrocyte membranes as they provide energy and neurotransmitters to local neuronal circuits and regulate their activity [50, 58]. Furthermore, these oscillations also correlate with cognitive performance as well as sleep patterns [31, 99].

These findings suggest that all brainwaves and bodily rhythms, such as the cardiorespiratory rhythm, baroreflex fluctuations, and oxyhemoglobin/deoxyhemoglobin fluctuations, etc., are correlated, particularly when calm and relaxed, which occurs with self-resonance [30, 33]. This means that information from one rhythm contains embedded information about other rhythms, albeit at varying levels of resolution. Therefore, each biofeedback modality can work through a similar mechanism of action to effect change, while the differences in intensities of the effects may be due to the level of resolution of the underlying master regulatory rhythm as conferred by the particular form or substrate of the feedback.

5. Using ILF neurofeedback to train the conductor of the brainwave symphony

Despite an attempt to present this information fairly and objectively, my personal bias for ILF neurofeedback is likely obvious. The way I describe ILF neurofeedback to my clients nowadays is to consider the brain like an orchestra and the brainwaves like a symphony, which, of course, is not my own original analogy [66]. Each brainwave is like the music playing from each section of the orchestra, such as the winds, brass, strings, or percussion sections. You can train each section separately, but, due to the nature of systems, by training - and possibly changing - one section, the other sections will also likely be perturbed in some way that may not be readily discernable or beneficial. Alternatively, you can train the conductor, which, in this analogy, is the ISO or ILF. Thus, when you train the ILF, you train the entire symphony of brainwaves, which continue to play together, but with greater harmony after neurofeedback training.

6. Conclusions

The future of neurofeedback depends on the diverse field of methods coming together and defining the mechanism(s) of action of neurofeedback that can be applied to all methods. Detailed mechanisms of action are likely different for the different methods, but their fundamental processes should not contradict each other. Not only will this help advance the field, but it will also help potential clients to understand how all neurofeedback works in general, then they can choose which methodology works best for them based on the specific characteristics of that method. A two-part mechanism of action is presented here, one that is general for all forms of biofeedback, and one that is more specific to neurofeedback, yet still general enough to be applicable to each specific method. Hopefully, other neurofeedback practitioners and researchers will consider these hypotheses, possibly further developing them and testing them through well-designed research studies. As the field and use of neurofeedback grows, these mechanistic models can be further refined to fit all methodologies and conditions.

Acknowledgements

The author acknowledges Susan Othmer, Clinical Director of the EEG Institute, Woodland, Hills, CA, for her methodical pursuit of developing the ILF neurofeedback technique and training thousands of clinicians from around the world in the method, and always being available to provide assistance and guidance. Siegfried and Kurt Othmer are also acknowledged for their theoretical perspectives on the underlying science as well as practical help in implementation of the neurofeedback business model. The author also acknowledges their father, James M. Markovics, for his constant support, in-depth theoretical discussions, and training assistance, which launched the author's second career as a neurofeedback practitioner.

Conflict of interest

The author declares no conflicts of interest.

Author details

Jen A. Markovics
Rose City Therapeutics, LLC., Tualatin, OR, United States

*Address all correspondence to: drjen@rosecitytherapeutics.com

References

[1] Teufel, C, Fletcher, PC. Forms of prediction in the nervous system. Nature Reviews Neuroscience. 2020;21;231-242. DOI: 10.1038/s41583-020-0275-5

[2] Greve, PF. The role of prediction in mental processing: A process approach. New Ideas in Psychology. 2015;39;45-52. DOI: 10.1016/j.newideapsych.2015.07.007

[3] Arnal, LH, Giraud, A-L. Cortical oscillations and sensory predictions. Trends in Cognitive Science. 2012;16(7);390-398. DOI: 10.1016/j.tics.2012.05.003

[4] Lehrer, P, Eddie, D. Dynamic processes in regulation and some implication for biofeedback and biobehavioral interventions. Applied Psychophysiology & Biofeedback. 2013;38(2);143-155. DOI: 10.1007/s10484-013-9217-6

[5] Ros, T, Baars, BJ, Lanius, RA, Vuilleumier, P. Tuning pathological brain oscillations with neurofeedback: a systems neuroscience framework. Frontiers in Human Neuroscience. 2014;8;1008. DOI: 10.3389/fnhum.2014.01008

[6] Hovseypyan, S, Olasagasti, I, & Giraud, A-L. Combining predictive coding and neural oscillations enables online syllable recognition in natural speech. Nature Communications. 2020;11;3117. DOI: 10.1038/s41467-020-16956-5

[7] Kropotov, J. (2016). Functional Neuromarkers of Psychiatry: Applications for Diagnosis and Treatment. San Francisco: Academic Press, 2016. 462 p. ISBN: 978-0-12-410513-3

[8] Luck, SJ, Kappenman, ES. The Oxford Handbook of Event-Related Potential Components. New York: Oxford University Press, 2011. 664 p. DOI: 10.1093/oxfordhb/9780195374148.001.0001

[9] McCaldon, RJ. Automatism. Canadian Medical Association Journal. 1964;91(17):914-920.

[10] Sarva, H, Deik, A, Severt, WL. Pathophysiology and treatment of alien hand syndrome. Tremor and Other Hyperkinetic Movements. 2014;4;241. DOI: 10.7916/D8VX0F48

[11] Chmielewski, WX, Zink, N, Chmielewski, KY, Beste, C, Stock, AK. How high-dose alcohol intoxication affects the interplay of automatic and controlled processes. Addiction biology. 2020;25(1);e12700. DOI: 10.1111/adb.12700

[12] Faber, R, Azad, A, Reinsvold, R. A case of the corpus callosum and alien hand syndrome from a discrete paracallosal lesion. Neurocase. 2010;16(4);281-285. DOI: 10.1080/13554790903456217

[13] Brugger, F, Galovic, M, Weder, BJ, Kägi, G. Supplementary motor complex and disturbed motor control – a retrospective clinical and lesion analysis of patients after anterior cerebral artery stroke. Frontiers in Neurology. 2015;6;209. DOI: 10.3389/fneur.2015.00209

[14] Hassan, A, Josephs, KA. Alien hand syndrome. Current Neurology and Neuroscience Reports. 2016;16(8). DOI: 10.1007/s11910-016-0676-z

[15] Law, CSH, Leung, LS. Long-term potentiation and excitability in the hippocampus are modulated differentially by theta rhythm. eNeuro. 2018;5(6);ENEURO.0236-18.2018. DOI: 10.1523/ENEURO.0236-18.2018

[16] Shulz, DE, Feldman, DE. Spike timing-dependent plasticity. In: Rubenstein, JLR, Rakic, P, editors. Neural Circuit Development and Function in the Brain. San Francisco: Academic Press; 2013. p. 155-181. DOI: 10.1016/B978-0-12-397267-5.00029-7

[17] Moulson, MC, Nelson, CA. Neurological development. In: Haith, MM, & J.B. Benson, JB, editors. Encyclopedia of Infant and Early Childhood Development. San Francisco: Academic Press; 2008. p. 414-424. DOI: 10.1016/b978-012370877-9.00109-2

[18] Perez-Catalan, NA, Doe, CQ, Ackerman, SD. The role of astrocyte-mediated plasticity in neural circuit development and function. Neural Development, 2021;16;1. https://doi.org/10.1186/s13064-020-00151-9

[19] Byrne, JH, Roberts, JL, editors. From Molecules to Networks: An Introduction to Cellular and Molecular Neuroscience. 2nd ed. San Francisco: Academic Press; 2009. ISBN: 978-0-12-374132-5.

[20] Fox, K, Stryker, M. Integrating Hebbian and homeostatic plasticity: introduction. Philosophical Transactions of the Royal Society B. 2017;372;20160413. DOI: 10.1098/rstb.2016.0413

[21] Frölich, F. Network Neuroscience. San Francisco: Academic Press; 2016. 482 p. DOI: 10.1016/C2013-0-23281-5

[22] Engel, AK, Gerloff, C, Hilgetag, CC, Nolte, G. Intrinsic coupling modes: Multiscale interactions in ongoing brain activity. Neuron. 2013;80;867-886. DOI: 10.1016/j.neuron.2013.09.038

[23] Korte, M., Schmitz, D. Cellular and systems biology of memory: Timing, molecules, and beyond. Physiological Review, 2016;96;647-693. DOI: 10.1152/physrev.00010.2015

[24] Lv, H, Wang, Z, Tong, E, Williams, LM, Zaharchuk, G, Zeineh, M, Goldstein-Piekarski, AN, Ball, TM, Liao, C, Wintermark, M. Resting-state functional MRI: Everything that nonexperts have always wanted to know. American Journal of Neuroradiology. 2018;39(8);1380-1399. DOI: 10.3174/ajnr.A5527

[25] Medaglia, J.D., Lynall, M.E., & Bassett, D.S. Cognitive network neuroscience. Journal of Cognitive Neuroscience, 2015;27(8);1471-1491. https://doi.org/10.1162/jocn_a_00810

[26] Raichle, ME. The Restless Brain. Brain Connectivity. 2011;1(1);3-12. DOI: 10.1089/brain.2011.0019

[27] Raichle, ME. The restless brain: how intrinsic activity organizes brain function. Philosophical Transactions of the Royal Society B. 2015;370;20140172. DOI: 10.1098/rstb.2014.0172

[28] Barttfeld, P, Uhrig, L, Sitt, JD, Sigman, M, Jarraya, B, Dahaene, S. Signature of consciousness in brain dynamics. Proceedings of the National Academy of Sciences. 2015;112(3):887-892. DOI: 10.1073/pnas.1418031112

[29] Chow, HM, Horovitz, SG, Carr, WS, Picchioni, D, Coddington, N, Fukunaga, M, Xu, Y, Balkin, TJ, Duyn, JH, Braun, AR. Rhythmic alternating patterns of brain activity distinguish rapid eye movement sleep from other states of consciousness. Proceedings of the National Academy of Sciences of the United States of America. 2013;110(25); 10300-10305. DOI: 10.1073/pnas.1217691110

[30] Jerath, R, Crawford, MW. Layers of human brain activity: A functional model based on the default mode network and slow oscillations. Frontiers in Human Neuroscience. 2015;9;248. DOI: 10.3389/fnhum.2015.00248

[31] Watson, BO. Cognitive and physiologic impacts of the infraslow

oscillation. Frontiers in Systems Neuroscience. 2018;12;44. DOI: 10.2889/fnsys.2018.00044

[32] Wen, X, Yao, L, Fan, T, Wu, X, Liu, J. The spatial pattern of basal ganglia network: A resting state fMRI study. In: Proceedings of 2012 International Conference on Complex Medical Engineering; 1-4 July 2012; Kobe, Japan. p. 43-46. DOI: 10.1109/ICCME.2012.6275632

[33] Jerath, R, Beveridge, C. Mysteries of the mind: Insights into the default space model of consciousness. Frontiers in Human Neuroscience. 2018;12;162. DOI: 10.3389/fnhum.2018.00162

[34] Ioannides, AA. Neurofeedback and the neural representation of self: lessons from awake state and sleep. Frontiers in Human Neuroscience. 2018;12;142. DOI: 10.2289/fnhum.2017.00142

[35] Di, X, Biswal, BB. Modulatory interactions of resting-state brain functional connectivity. PLoS ONE. 2013;8(8);e71163. DOI: 10.1371/journal.pone.0071163

[36] Baker, AP, Brookes, MJ, Rezek, IA, Smith, SM, Behrens, T, Probert Smith, PJ, Woolrich, M. Fast transient networks in spontaneous human brain activity. eLife. 2014;3;e01867. DOI: 10.7554/eLife.01867

[37] Di, X, Biswal, BB. Modulatory interactions between the default mode network and task positive networks in resting-state. PeerJ. 2014;2;e367. DOI: 10.7717/peerj.367

[38] Zhang, D, Liang, B, Wu, X, Wang, Z, Xu, P, Chang, S, Liu, B, Liu, M, Huang, R. Directionality of large-scale resting-state brain networks during eyes open and eyes closed conditions. Frontiers in Human Neuroscience. 2015;9;81. DOI: 10.3389/fnhum.2015.00081

[39] Uddin LQ. Cognitive and behavioural flexibility: neural mechanisms and clinical considerations. Nature Reviews Neuroscience. 2021;22(3);167-179. DOI: 10.1038/s41583-021-00428-w

[40] Cohen, JR. The behavioral and cognitive relevance of time-varying, dynamic changes in functional connectivity. Neuroimage. 2018;180 (Pt B);515-525. DOI: 10.1016/j.neuroimage.2017.09.036

[41] Durstewitz, D, Huys, QJM, Koppe, G. Psychiatric illnesses as disorders of network dynamics. Biological Psychiatry: Cognitive Neuroscience and Neuroimaging, 2020. DOI: 10.1016/j.bpsc.2020.01.001

[42] Jia, H, Hu, X, Deshpande, G. Behavioral relevance of the dynamics of the functional brain connectome. Brain connectivity. 2014;4(9);741-759. DOI: 10.1089/brain.2014.0300

[43] Rashid, B, Damaraju, E, Pearlson, GD, Calhoun, VD. Dynamic connectivity states estimated from resting fMRI identify differences among Schizophrenia, bipolar disorder, and healthy control subjects. Frontiers in Human Neuroscience. 2014;8;897. DOI: 10.3389/fnhum.2014.00897

[44] Schaefer A, Margulies, DS, Lohmann, G, Gorgolewski, KJ, Smallwood, J, Kiebel, SJ, Villringer, A. Dynamic network participation of functional connectivity hubs assessed by resting-state fMRI. Frontiers in Human Neuroscience. 2014;8;195. DOI: 10.3389/fnhum.2014.00195

[45] Escrichs A, Biarnes C, Garre-Olmo J, Fernández-Real JM, Ramos R, Pamplona R, Brugada R, Serena J, Ramió-Torrentà L, Coll-De-Tuero G, Gallart L, Barretina J, Vilanova JC, Mayneris-Perxachs J, Essig M, Figley CR, Pedraza S, Puig J, Deco G. Whole-Brain Dynamics in

Aging: Disruptions in Functional Connectivity and the Role of the Rich Club. Cerebral Cortex. 2021;31(5):3466-2481. doi:10.1093/cercor/bhaa367

[46] Song J, Birn RM, Boly M, Meier TB, Nair VA, Meyerand ME, Prabhakaran V. Age-related reorganizational changes in modularity and functional connectivity of human brain networks. Brain Connect. 2014;4(9);662-76. DOI: 10.1089/brain.2014.0286

[47] Qin, J, Chen, S-G, Hu, D, Zeng, L-L, Fan, Y-M, Chen, X-P, Shen, H. Predicting individual brain maturity using dynamic functional connectivity. Frontiers in Human Neuroscience. 2015;9;418. DOI: 10.3389/fnhum.2015.00418

[48] Kaiser, DA. The role of glia and astrocytes in brain functioning. In: Kirk, HW, editor. Restoring the Brain: Neurofeedback as an Integrative Approach to Health. Boca Raton, FL: CRC Press; 2016. p. 51-58. ISBN: 978-1-4822-5877-6

[49] Maynard, RL, Downes, N. The Brain and Spinal Cord. In: Anatomy and Histology of the Laboratory Rat in Toxicology and Biomedical Research. San Francisco: Academic Press; 2019. p. 231-260. ISBN: 978-0-12-811837-5

[50] Sardar, D, Cheng, Y-T, Szewcyk, LM, Deneen, B, Molofsky, AV. Mechanisms of astrocyte development. In: Rubenstein, J, Rakic, P, Chen, S, senior editors. Patterning and Cell Type Specification in the Developing CNS and PNS. 2nd ed. San Francisco: Academic Press; 2020. DOI: 10.1016/B978-0-12-814405-3.00032-1

[51] Verkhratsky, A, Butt, AM. The history of the decline and fall of the glial numbers legend. Neuroglia. 2018;1;188-192. DOI: 10.3390/neuroglia1010013

[52] Mederos, S, González-Arias, C, Perea, G. Astrocyte-neuron networks: A multilane highway of signaling for homeostatic brain function. Frontiers in Synaptic Neuroscience. 2018;10;45. DOI: 10.3389/fnsyn.2018.00045

[53] Spampinato, SF, Bortolotto, V, Canonico, PL, Sortino, MA, Grilli, M. Astrocyte-derived paracrine signals: Relevance for neurogenic niche regulation and blood-brain barrier integrity. Frontiers in Pharmacology. 2019;10;1346. DOI: 10.3389/fphar.2019.01346

[54] Simard, M, Arcuino, G, Takano, T, Liu, QS, Nedergaard, M. Signaling at the gliovascular interface. The Journal of Neuroscience. 2003;23(27);9254-0262.

[55] Huang, AY, Woo, J, Sardar, D, Lozzi, B, Bosquez Huerta, NA, Lin, CJ, Felice, D, Jain, A, Paulucci-Holthauzen, A, Deneen, B. Region-specific transcriptional control of astrocyte function oversees local circuit activities. Neuron 2020;106(6);992-1008.e9. DOI: 10.1016/j.neuron.2020.03.025

[56] Hwang, SN, Lee, JS, Seo, K, Lee, H. Astrocytic regulation of neural circuits underlying behaviors. Cells. 2021;10(2); 296. DOI: 10.3390/cells10020296

[57] Buskila, Y, Bllot-Saez, A, Morley, JW. Generating brain waves, the power of astrocytes. Frontiers in Neuroscience. 2019;13;1125. DOI: 10.3389/fnins.2019.01125

[58] Martinez-Banaclocha, M. Astroglial isopotentiality and calcium-associated biomagnetic field effects on cortical neuronal coupling. Cells. 2020;9;439. DOI: 10.3390/cells9020439

[59] Najjar, S, Pearlman, DM, Alper, K, Najjar, A, Devinsky, O. Neuroinflammation and psychiatric illness. Journal of Neuroinglammation. 2013;10;43. DOI: 10.1186/1742-2094-10-43

[60] Larsen, S. The Neurofeedback Solution: How to Treat Autism, ADHD,

Anxiety, Brain Injury, Stroke, PTSD, and More. Fairfield, CT: Healing Arts Press; 2012. 424 p. ISBN: 978-1-59477-366-2

[61] Marzbani, H, Marateb, HR, Mansourian, M. Neurofeedback: A comprehensive review on system design, methodology and clinical applications. Basic and Clinical Neuroscience, 2016;7(2);143-158. DOI: 10.15412/J.BCN.03070208

[62] Micoulaud-Franchi, J-A, McGonigal, A, Lopez, R, Daudet, C, Kotwas, I, Bartolomei, F. Electroencephalographic neurofeedback: Level of evidence in mental and brain disorders and suggestions for good clinical practice. Clinical Neurophysiology. 2015;45; 423-433.

[63] Niv, S. Clinical efficacy and potential mechanisms of neurofeedback. Personality and Individual Differences. 2013;54(6);676-686. DOI: 10.1016/j.paid.2012.11.037

[64] Omejc, N, Rojc, B, Battaglini, PP, & Marusic, U. Review of the therapeutic neurofeedback method using electroencephalography: EEG Neurofeedback. Bosnian Journal of Basic Medical Sciences. 2018;19(3);213-220. DOI: 10.17305/bjbms.2018.3785

[65] Othmer, S. History of neurofeedback In: Kirk, HW, editor. Restoring the Brain: Neurofeedback as an Integrative Approach to Health. Boca Raton, FL: CRC Press; 2016. p. 23-50. ISBN: 978-1-4822-5877-6

[66] Robbins, J. A Symphony in the Brain: The Evolution of the New Brain Wave Biofeedback. New York: Grove Press; 2008. 272 p. ISBN: 978-0802143815

[67] Sterman, MB, LoPresti, RW, Fairchild, MD. Electroencephalographic and behavioral studies of monomethyl hydrazine toxicity in the cat. Journal of Neurotherapy: Investigations in Neuromodulation, Neurofeedback and Applied Neuroscience. 2010;14(4);293-300. DOI: 10.1080/10874208.2010.523367

[68] Nigro, SE. The efficacy of neurofeedback for pediatric epilepsy. Applied Psychophysiology and Biofeedback. 2019;44(4);285-290. DOI: 10.1007/s10484-019-09446-y

[69] Walker, JE., Kozlowski, GP. Neurofeedback treatment of epilepsy. Child and adolescent psychiatric clinics of North America. 2005;14(1);163–viii. DOI: 10.1016/j.chc.2004.07.009

[70] American Academy of Pediatrics. Evidence-based child and adolescent psychosocial interventions. Itasca, IL: American Academy of Pediatrics; 2013.

[71] Cueli, M, Rodríguez, C, Cabaleiro, P, García, T, González-Castro, P. Differential efficacy of neurofeedback in children with ADHD presentations. Journal of Clinical Medicine. 2019;8;204. DOI: 10.3390/jcm8020204

[72] Enriquez-Geppert, S, Smit, D, Pimenta, MG, Arns, M. Neurofeedback a treatment intervention in ADHD: Current evidence and practice. Current Psychiatry Reports. 2019;21;46. DOI: 10.1007/s11920-019-1021-4

[73] Pigott, HE, Cannon, R. Neurofeedback is the best available first-line treatment for ADHD: What is the evidence for this claim? NeuroRegulation. 2014;1(1);4-23. DOI: 10.15540/nr.1.1.4

[74] Legarda, SB, McMahon, D, Othmer, S, Othmer, S. Clinical neurofeedback: Case Studies, proposed mechanism, and implicatios for pediatric neurology practice. Journal of Child Neurology. 2011;26(8);1045-1051. DOI: 10.1177/0883073811405052

[75] Orndorff-Plunkett, F, Singh, F, Aragón, OR, Pineda, JA. Assessing the effectiveness of neurofeedback training in the context of clinical and social neuroscience. Brain Sciences. 2017;7;95. DOI: 10.3390/brainsci7080095

[76] Othmer, S, Othmer, S. Infra-low frequency neurofeedback for optimum performance. Biofeedback. 2016;44(2):81-89. DOI: 10.5298/1081-5937-44.2.07

[77] Brenninkmeijer, J. Brainwaves and psyches. History of the Human Sciences. 2015;28(3);115-133. doi:10.1177/0952695114566644

[78] Masterpasqua, F, Healey, KN. Neurofeedback in Psychological Practice. Professional Psychology: Research and Practice. 2003;34(6);652-656. doi:10.1037/0735-7028.34.6.652

[79] Herrmann, CS, Strüber, D, Helfrich, RF, Engel, AK. EEG oscillations: From correlation to causality. International Journal of Psychophysiology. 2015;103;12-21. DOI: 10.1016/j.ijpsycho.2015.02.003

[80] Idris, Z, Muzaimi, M, Ghani, R, Idris, B, Kandasamy, R, &Abdullah, J. Principles, anatomical origin and applications of brainwaves: a review, our experience and hypothesis related to microgravity and the question on soul. Journal of Biomedical Science and Engineering. 2014;7;435-445. DOI: 10.4236/jbise.2014.78046

[81] Stern, JM. Atlas of EEG Patterns. 2nd ed. Philadelphia: Lippincott, Williams & Wilkins; 2013. ISBN: 978-1451109634

[82] Knyazev, GG. EEG delta oscillations as a correlate of basic homeostatic and motivational processes. Neuroscience and Biobehavioral Reviews. 2012;36;677-695. DOI: 10.1016/j.neurobiorev.2011.10.002

[83] Hipp, JF, Engel, AK, Siegel, M. Oscillatory synchronization in large-scale cortical networks predicts perception. Neuron. 2011;69(2);387-396. DOI: 10.1016/j.neuron.2010.12.027

[84] Muñoz-Moldes, S, Cleeremans, A. Delineating implicit and explicit processes in neurofeedback learning. Neuroscience and Biobehavioral Reviews. 2020;118;681-688. DOI: 10.1016/j.neubiorev.2020.09.003

[85] Dobrushina, OR, Vlasova, RM, Rumshiskaya, AD, Litvinova, LD, Mershina, EA, Sinitsyn, VE, Pechenkova, EV. Modulation of intrinsic brain connectivity by implicit electroencephalographic neurofeedback. Frontiers in Human Neuroscience. 2020;14;192. DOI: 10.3389/fnhum.2020.00192

[86] Weber, LA, Ethofer, T, Ehlis, A-C. Predictors of neurofeedback training outcome: A systematic review. Neuroimage. 2020;27;102301. DOI: 10.1016/j.psychres.2011.12.041

[87] Goldstein, EB. Cognitive Psychology: Connecting Mind, Research and Everyday Experience. 4th ed. Stamford, CT: Cengage Learning; 2014. 464 p. ISBN: 978-1285763880

[88] Coben, R, Hammond, DC, Arns, M. 19 channel z-score and LORETA neurofeedback: Does the evidence support the hype? Applied Psychophysiology and Biofeedback. 2019;44;1-8. DOI: 10.1007/s10484-018-9420-6

[89] Simkin, DR, Thatcher, RW, Lubar, J. Quantitative EEG and neurofeedback in children and adolescents: Anxiety disorders, depressive disorders, addiction and attention-deficit/hyperactivity disorder, and brain injury. Child and Adolescent Psychiatric Clinics of North America. 2014;23(3);427-464. DOI: 10.1016/j.chc.2014.03.001

[90] Koberda, JL. Z-score LORETA Neurofeedback as a Potential Therapy in Depression/Anxiety and Cognitive Dysfunction. In: Lubar, JF, Thatcher, RW, editors. Z Score Neurofeedback. San Diego: Elsevier Science; 2015. p. 93-113. DOI: 10.1016/b978-0-12-801291-8.00005-4

[91] Dudek, E, Dodell-Feder, D. The efficacy of real-time functional magnetic resonance imaging neurofeedback for psychiatric illness: A meta-analysis of brain and behavioral outcomes. Neuroscience and biobehavioral reviews. 2021;121;291-306. DOI: 10.1016/j.neubiorev.2020.12.020

[92] Emmert, K, Kopel, R, Sulzer, J, Brühl, AB, Berman, BD, Linden, D, Horovitz, SG, Breimhorst, M, Caria, A, Frank, S, Johnston, S, Long, Z, Paret, C, Robineau, F, Veit, R, Bartsch, A, Beckmann, CF, Van De Ville, D, Haller, S. Meta-analysis of real-time fMRI neurofeedback studies using individual participant data: How is brain regulation mediated? NeuroImage. 2016;124(Pt A);806-812. DOI: 10.1016/j.neuroimage.2015.09.042

[93] Thibault, RT, MacPherson, A, Lifshitz, M, Roth, RR, Raz, A. Neurofeedback with fMRI: A critical systematic review. NeuroImage. 2018;172;786-807. DOI: 10.1016/j.neuroimage.2017.12.071

[94] Gomes, JS, Ducos, DV, Gadelha, A, Ortiz, BB, Van Deusen, AM, Akiba, HT, Guimaraes, L, Cordeiro, Q, Trevizol, AP, Lacerda, A, Dias, AM. Hemoencephalography self-regulation training and its impact on cognition: A study with schizophrenia and healthy participants. Schizophrenia Research. 2018;195;591-593. DOI: 10.1016/j.schres.2017.08.044

[95] Refinetti, R. Integration of biological clocks and rhythms. Comprehensive Physiology. 2012;2;1213-1239. DOI: 10.1002/cphy.c100088

[96] Llewellyn Smith, M, Collura, TF, Ferrera, J, de Vries, J. Infra-slow fluctuation training in clinical practice: A technical history. NeuroRegulation. 2014;1(2);187-207. DOI: 10.15540/nr.1.2.187

[97] Othmer, S, Othmer, SF, Kaiser, DA, Putnam, J. Endogenous neuromodulation at infra-low frequencies. Seminars in Pediatric Neurology. 2013;20(4);246-257 DOI: 10.1016/j.spen.2013.10.006

[98] Grin-Yatsenko, V, Kara, O, Evdokimov, SA, Gregory, M, Othmer, S, Kropotov, JD. Infra-low frequency neuro feedback modulates infra-slow oscillations of brain potentials: a controlled study. Journal of Biomedical Engineering and Research. 2020;4;1-11. DOI: 10.5772/intechopen.77154

[99] Monto, S, Palva, S, Voipio, J, Palva, JM. Very slow EEG fluctuations predict the dynamics of stimulus detection and oscillation amplitudes in humans. The Journal of Neuroscience. 2008;28(33);8268-8272. DOI: 10.1523/JNEUROSCI.1910-08.2008

[100] Othmer, S. (2019). EEGInfo Protocol Guide for Neurofeedback Clinicians. 7th ed. Woodland Hills, CA: EEG Info; 2019. 166 p. ISBN: 978-0-9895432-7-9

[101] Wiedemann, M. The evolution of clinical neurofeedback practice. In: Kirk, HW, editor. Restoring the Brain: Neurofeedback as an Integrative Approach to Health. Boca Raton, FL: CRC Press; 2016. p. 59-91. ISBN: 978-1-4822-5877-6

[102] NeurOptimal®. Discover NeurOptimal® [Internet]. 2021. Available from: https://neuroptimal.com/discover-neuroptimal-usa/

[103] Alvarez, J, Meyer, FL, Granoff, DL, Lundy, A. The effect of EEG biofeedback on reducing postcancer cognitive impairment. Integrative cancer therapies, 2013;12(6);475-487. DOI: 10.1177/1534735413477192

[104] Harris, S, Lambie, GW, Hundley, G. The effects of neurofeedback training on college students' attention deficit hyperactivity disorder symptoms. Counseling Outcome Research and Evaluation, 2018;1-14. doi:10.1080/21501378.2018.1442679

[105] Brown, VW. Neurofeedback and Lyme's disease. Journal of Neurotherapy. 1995;1(2);60-73. DOI: 10.1300/j184v01n02_05

[106] Jackson, M. "Divine Stramonium": The rise and fall of smoking for asthma. Medical History, 2010;54;171-194.

[107] Kohl, SH, Mehler, DMA, Lührs, M, Thibault, RT, Konrad, K, Sorger, B. The potential of functional near-infrared spectroscopy-based neurofeedback – A systematic review and recommendations for best practice. Frontiers in Neuroscience. 2020;14;594. DOI: 10.3389/fnins.2020.00594

[108] Steriade, M, Gloor, P, Llinas, RR, Lopes de Silva, FH, Mesulam, M-M. Basic mechanisms of cerebral rhythmic activities. Electroencephalography and Clinical Neurophysciology. 1990;76;481-508.

[109] Newson, JJ, Thiagarajan, TC. EEG frequency bands in psychiatric disorders: A review of resting state studies. Frontiers in Human Neuro science. 2019;12;521. DOI: 10.33889/fnhum.2018.00521

[110] Ogrim, G, Kropotov, J, Hestad, K. The QEEG theta/beta ratio in ADHD and normal controls: Sensitivity, specificity, and behavioral correlates. Psychiatry Research. 2012;198(3);482-488. DOI: 10.1016/j.psychres.2011.12.041

[111] Arns, M, Conners, CK, Kraemer, HC. A decade of EEG theta/beta ratio research in ADHD: A meta-analysis. Journal of Attention Disorders. 2011;17(5);374-383. DOI: 10.1177/1087054712460087

[112] Snyder, SS, Rugino, TA, Homig, M, Stein, MA. Integration of an EEG biomarker with a clinician's ADHD evaluation. Brain and Behavior. 2015;0(0);e00330. DOI: 10.1002/brb3.330

[113] Janssen, TWP, Blink, M, Weeda, WD, Geladé, K, van Mourik, R, Maras, A, Oosterlaan, J. Learning curves of theta/beta neurofeedback in children with ADHD. European Child & Adolescent Psychiatry. 2017;26;573-582. DOI: 10.1007/s00787-016-0920-8

[114] Zuberer, A, Brandeis, D, Drechsler, R. Are treatment effects of neurofeedback training in children with ADHD related to the successful regulation of brain activity? A review on the learning of regulation of brain activity and a contribution to the discussion on specificity. Frontiers in human neuroscience. 2015;9;135. DOI: 10.3389/fnhum.2015.00135

[115] Othmer, S, Othmer, S, Legarda, S. Clinical neurofeedback: Training brain behavior. Treatment Strategies-Pediatric Neurology and Psychiatry. 2011;2;67-73.

[116] Jansen, R, Han, LKM, Verhoeven, JE, Aberg, KA, van den Oord, ECGJ, Milaneschi, Y, Penninx, BWJH. An integrative study of five biological clocks in somatic and mental health. eLife. 2021;10;59479. DOI: 10.7554/eLife.59479

[117] Porges, SW. The polyvagal theory: New insights into adaptive reactions of the autonomic nervous system. Cleveland Clinic Journal of Medicine.

2009;76(Suppl 2);S86-S90. DOI: 10.3949/ccjm.76.s2.17

[118] Huttunen, MO, Mednick, SA. Polyvagal theory, neurodevelopment and psychiatric disorders. Irish Journal of Psychological Medicine. 2018;35;9-10. DOI: 10.1017/ipm.2017.66

[119] Porges SW. The polyvagal perspective. Biological psychology. 2007;74(2);116-143. DOI: 10.1016/j.biopsycho.2006.06.009

[120] Hinterberger, T, Walter, N, Doliwa, C, Loew, T. The brain's resonance with breathing - decelerated breathing synchronizes heart rate and slow cortical potentials. Journal of Breath Research. 2019;13(4);046003. DOI: 10.1088/1752-7163/ab20b2

[121] Imperatori, C, Della Marca, G, Amoroso, N, Maestoso, G, Valenti, EM, Massullo, C, Carbone, GA, Contardi, A, Farina, B. Alpha/theta neurofeedback increases mentalization and the default mode network connectivity in a non-clinical sample. Brain Topography. 2017;30(6);822-831. DOI: 10.1007/s10548-017-0593-8

[122] Russell-Chapin, L, Kemmerly, T, Liu, W-C, Zagardo, MT, Chapin, T, Daily, D, Dinh, D. The effects of neurofeedback in the default mode network: Pilot study results of medicated children with ADHD. Journal of Neurotherapy: Investigations in Neuromodeulation, Neurofeedback and Applied Neuroscience. 2013;17(1);35-42. DOI: 10.1080/10874208.2013.759017

[123] Yamashita, A, Hayasaka, S, Kawato, M, Imamizu, H. Connectivity neurofeedback training can differentially change functional connectivity and cognitive performance. Cerebral Cortex. 2017;27;4960-4970. DOI: 10.1093/cercor/bhx177

Chapter 7

Entropy and the Emotional Brain: Overview of a Research Field

Beatriz García-Martínez, Antonio Fernández-Caballero and Arturo Martínez-Rodrigo

Abstract

During the last years, there has been a notable increase in the number of studies focused on the assessment of brain dynamics for the recognition of emotional states by means of nonlinear methodologies. More precisely, different entropy metrics have been applied for the analysis of electroencephalographic recordings for the detection of emotions. In this sense, regularity-based entropy metrics, symbolic predictability-based entropy indices, and different multiscale and multilag variants of the aforementioned methods have been successfully tested in a series of studies for emotion recognition from the EEG recording. This chapter aims to unify all those contributions to this scientific area, summarizing the main discoverings recently achieved in this research field.

Keywords: Emotion recognition, Electroencephalographic recordings, Entropy metrics, Nonlinear analysis, Survey

1. Introduction

Emotions are essential in our daily lives, with an enormous repercussion on perception, cognition, learning and rational decision-making processes [1]. As a result, the affective neuroscience has emerged with the purpose of studying the influence of emotions on areas like psychology, philosophy or neurobiology, among many others [1]. The emotional states defined in the literature range from a few basic emotions [2] to several complex emotions created as combinations of the basics [3]. These emotional states can be classified according to different models, being the circumplex model of Russell one of the most widely used [4]. This bidimensional model distributes all the existing emotional states according to two emotional parameters, namely valence and arousal. Valence represents the degree of pleasantness or unpleasantness produced by an emotional stimulus, whereas arousal measures the activation or deactivation that a stimulus provokes. The location of each emotional state in the circumplex model is determined by its level of both dimensions, as shown in **Figure 1**.

Emotions also play a key role in human communication and interaction processes. Nevertheless, the human-machine interfaces (HMIs) are still not able to identify human emotional states. In a digital society in which those systems are daily introduced in multiple ordinary scenarios, it becomes crucial to supplement this lack of emotional intelligence of HMIs. In this sense, the aim of the Affective Computing science is to endow those systems with the capability to automatically

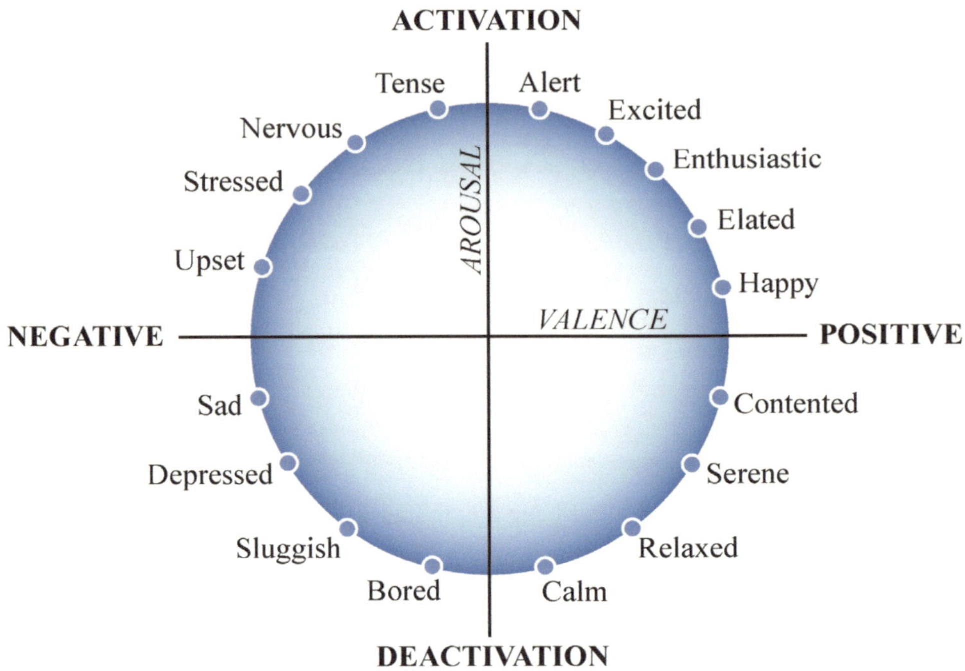

Figure 1.
Circumplex model of Russell for classification of emotions based on their level of valence and arousal.

detect and interpret human emotions and decide which actions to execute accordingly, thus improving the interactions between people and machines [5, 6].

The detection of emotional states can be conducted by means of the assessment of bodily reactions to emotional stimuli, for which different physiological variables can be measured and analyzed. One of the most widely studied in the last years is the electroencephalography (EEG), which represents the electrical activity generated in the brain due to neural connections [7]. The selection of EEG recordings instead of other physiological signals is justified by the fact that the brain generates the first impulse against any stimulus, and then it is spread to the rest of peripheral systems through the central nervous system. In this sense, EEG signals represent the activity of the source of the emotional response, whereas the rest of physiological variables can be considered as secondary effects of the brain's performance [8]. As a consequence, the number of works focused on the analysis of EEG time series for emotions detection has notably increased in the last years [9].

The evaluation of EEG recordings has been traditionally conducted from a linear perspective, especially in the frequency domain, studying features such as the spectral power or the asymmetry between the two brain hemispheres in different frequency bands [10]. However, the brain activity is far from being considered linear. Contrarily, neural processes follow a completely heterogeneous and nonstationary performance even at both cellular and global level [11]. With this respect, the application of linear algorithms may not report a complete description of the brain's behavior [12]. For this reason, nonlinear methodologies have been widely applied for discovering underlying information unrevealed by traditional linear techniques [13]. Indeed, nonlinear indices have already outperformed the results derived from the application of those linear algorithms for the evaluation of various mental processes, including the recognition of emotions [13].

Among the different nonlinear methodologies that can be found in the literature, entropy indices have been widely applied in the context of emotions recognition with EEG recordings [14]. Entropy represents the rate of information reported by a time series, describing the nonlinear characteristics of a nonstationary system [15]. Hence, entropy metrics become promising tools for the assessment of the chaotic dynamics of a nonstationary system such as the brain. Indeed, in the literature many studies have applied these nonlinear methodologies for the identification of emotional states from EEG recordings. The present manuscript summarizes the main discoverings of the last years in the scientific field of emotions recognition from EEG signals with entropy indices.

2. Entropy indices

Entropy was firstly defined in thermodynamics, referring to the distribution probability of molecules in a fluid system [15]. In information theory and signals analysis, this concept was adapted by Shannon, who defined entropy as a measure of the information provided by a time series, describing its complexity, irregularity or unpredictability [16]. With respect to the EEG analysis, many entropy indices have been introduced and successfully applied for the study of various physical and mental disorders, like epilepsy [17], Alzheimer [18], autism [19] or depression [20], among others. As a result of the valuable outcomes, entropy metrics have also been introduced in the research field of emotions recognition from EEG recordings [14]. The following subsections give a brief mathematical description of the entropy metrics mainly applied for emotions detection.

2.1 Regularity-based entropy indices

The irregularity of a signal represents the rate of repetitiveness of patterns, reaching higher values for non-repetitive and disordered time series, and lower for sequences with a high rate of occurrence [21]. One of the regularity-based entropy metrics widely used is the approximate entropy (ApEn), which evaluates the probability of having repetitive patterns and assigns a non-negative number to each sequence in terms of its repetitiveness, with lower values for more recurrent patterns [22]. Mathematically, ApEn is computed as

$$\mathrm{ApEn}(m, r) = C^{m}(r) - C^{m+1}(r), \tag{1}$$

where $C^{m}(r)$ and $C^{m+1}(r)$ are the correlation integrals that represent the likelihood of having two sequences matching for m and for $m+1$ points, respectively, within the threshold r [22]. Nevertheless, ApEn also considers the self-matching of each sequence, thus influencing on the final result. Therefore, the sample entropy (SampEn) was developed to address this issue [23]. SampEn eliminates the self-matching and makes results independent of the value of the vector length m selected. If $B^{m}(r)$ is the probability that two patterns match for m points, and $B^{m+1}(r)$ is the probability that two patterns match for $m+1$ points, defined to exclude self-matches, then SampEn is calculated as

$$\mathrm{SampEn}(m, r) = -\ln\left[\frac{B^{m+1}(r)}{B^{m}(r)}\right]. \tag{2}$$

Moreover, quadratic sample entropy (QSampEn) emerged as an improvement of SampEn to make it insensitive to the value of the threshold r chosen for its calculation [24]. This independence of the parameter r is achieved by simply adding the term $\ln(2r)$ to the SampEn equation:

$$\text{QSampEn}(m, r) = \text{SampEn}(m, r, N) + \ln(2r). \tag{3}$$

2.2 Predictability-based and symbolic entropy indices

The predictability of a nonstationary system is related to its stable and deterministic evolution in time. Most of the entropy metrics for predictability measurement are symbolic indices that convert the original signal into a sequence of discrete symbols to form sequences [25]. After this symbolization, the evaluation of the predictability of a time series can be carried out with multiple techniques. The most commonly used is the Shannon entropy (ShEn), which quantifies the predictability of a signal in terms of the probability distribution of its amplitudes [16]. The mathematical expression of ShEn is

$$\text{ShEn}(m) = -\sum_{i=1}^{m} p(x_i) \cdot \ln(p(x_i)), \tag{4}$$

being $p(x_i)$ the probability of appearance of each symbolic sequence x_i of length m.

The Rényi entropy (REn) is a generalization of ShEn that is also widely used for the quantification of underlying dynamics in symbolized signals [26]. Concretely, REn provides a better characterization of some rare and frequent ordinal sequences, and it is defined as

$$\text{REn}(m, q) = -\frac{1}{1-q} \ln \sum_{i=1}^{m} p(x_i)^q, \tag{5}$$

being q ($q \geq 0$ and $q \neq 1$) the bias parameter that enables a more accurate characterization of a nonlinear signal [26]. Indeed, ShEn is the particular case of REn for $q = 1$, thus being REn a more flexible index than ShEn [26].

The version of ShEn for continuous random variables, called differential entropy (DEn), has received growing interest in the last years [27]. This entropy index can be expressed as

$$\text{DEn}(X) = -\int_X f(x) \log(f(x)) dx \tag{6}$$

where X is a random signal and $f(x)$ is its probability density function. In the case of time series governed by the Gauss distribution $N(\mu, \sigma^2)$, being μ and σ its mean and variance, respectively, DEn can be defined as

$$\text{DEn}(X) = -\int_{-\infty}^{\infty} \frac{1}{\sqrt{2\pi\sigma^2}} e^{-\frac{(x-\mu)^2}{2\sigma^2}} \log\left(\frac{1}{\sqrt{2\pi\sigma^2}} e^{-\frac{(x-\mu)^2}{2\sigma^2}}\right) dx = \frac{1}{2} \log(2\pi e\sigma^2). \tag{7}$$

As EEG signals follow a Gaussian distribution after the application of a band-pass filtering approach, the DEn of each sub-frequency band previously obtained by a Fast Fourier transform can be obtained according to the aforementioned equation [27].

Another widely used predictability-based entropy metric is permutation entropy (PerEn), which is a fast and insensitive to noise metric that evaluates the order of the symbols within a pattern [28]. Briefly, the original time series is symbolized to obtain ordinal sequences x_i that are associated with $m!$ permutation patterns κ_i. Considering $p(\pi_k)$ as the probability of apperarance of each permutation pattern, PerEn can be then computed by means of ShEn:

$$\text{PerEn}(m) = -\frac{1}{\ln(m!)}\sum_{k=1}^{m!} p(\pi_k) \cdot \ln(p(\pi_k)). \tag{8}$$

One of the limitations of PerEn is that it only considers the order of the symbols in a pattern, without taking into account their amplitudes. This limitation has been recently solved by means of the introduction of amplitude-aware permutation entropy (AAPE) [29]. This improvement of PerEn computes the probability $p^*(\pi_k)$ of appearance of patterns evaluating the average absolute and relative amplitudes of the symbolic sequences, and applying an adjustment coefficient to weight those parameters. Finally, AAPE is calculated in a similar manner as PerEn [29]:

$$\text{AAPE}(m) = -\frac{1}{\ln(m!)}\sum_{k=1}^{m!} p^*(\pi_k) \cdot \ln(p^*(\pi_k)). \tag{9}$$

Another option for the assessment of predictability of a time series is the spectral entropy (SpEn) [30]. In this case, the spectral power of a determined frequency is computed and normalized with respect to the total power, which gives a probability density function p_f. SpEn is then computed through ShEn or REn [30].

2.3 Multilag and multiscale entropy indices

The time series generated by nonlinear and nonstationary systems like the brain usually present highly complex dynamics derived from different simultaneous mechanisms that operate in multiple time scales [31]. As a result, the brain behavior cannot be completely described by means of single-scale methods. Therefore, multiscale variations of the aforementioned entropy metrics have been introduced with the purpose of revealing undiscovered information related to the multiscale nature of the EEG recordings. For the computation of multiscale entropy (MSE), the original signal $x(n)$ is firstly decomposed into coarse-grained time series $y^{(\kappa)}$, with κ as the scale factor, as follows:

$$y^{(\kappa)} = \frac{1}{\kappa}\sum_{i=(j-1)\kappa+1}^{j\kappa} x(i), \quad 1 \le j \le \frac{N}{\kappa}. \tag{10}$$

Therefore, all the previously defined entropy indices can be computed in a multiscale form for the coarse-grained series as defined above.

Another multiscale option is the wavelet entropy (WEn), which makes use of the decomposition of the original signal in different scales by means of the wavelet transform [32]. After the decomposition of the time series, the probability distribution p_j of the energy at each decomposition level is computed. Finally, WEn is estimated through ShEn:

$$\text{WEn} = -\sum_{j=1}^{q} p_j \cdot \ln\left(p_j\right). \tag{11}$$

On the other hand, the characteristics of the autocorrelation function of some signals require the consideration of a lag or time delay τ for a correct quantification of the complexity and nonlinear dynamics of the time series. In this sense, multilag entropy approaches are helpful to reduce the influence of the autocorrelation function for a proper characterization of a nonlinear signal [33]. One of those multilag approaches is the permutation min-entropy (PerMin), which is a symbolic time-delayed improvement of PerEn [34]. Starting from a generalization of PerEn through replacing ShEn by REn, the Rényi permutation entropy (RPE) is obtained as

$$\mathrm{RPE}(m, q, \tau) = \frac{1}{\ln(m!)} \cdot \frac{1}{1-q} \ln\left(\sum_{k=1}^{m!} p^{\tau}(\pi_k)^q\right). \tag{12}$$

Finally, PerMin is obtained in the limit $q \to \infty$ and presents the following expression:

$$\mathrm{PerMin}(m, \tau) = -\frac{1}{\ln(m!)} \ln\left(\max_{k=1,2,\dots,m!} [p^{\tau}(\pi_k)]\right). \tag{13}$$

3. Literature overview

In the literature, there are various studies that have applied regularity-based entropy measures for emotions detection with EEG recordings. A brief summary of those works is presented in **Table 1**, with information about the year of publication, the experimental design (including the number of emotions detected, subjects, EEG channels and type of stimulus), the features extracted, the classification models implemented, and the results obtained in each case. As can be observed, the interest in these metrics started growing in 2014, especially computing ApEn and SampEn for the detection of a number of emotions ranging between 2 and 4. In many cases, the signals analyzed were extracted from the publicly available Database for Emotion Analysis using Physiological Signals (DEAP), which consisted on a total of 32 healthy subjects watching emotional videoclips during the registration of their EEG with 32 channels [47]. Hence, different studies tested their methods on the same EEG recordings, thus allowing a direct comparison of the results obtained [37–39, 41, 42, 48]. The rest of works had different experimental designs. In terms of the classification models, support vector machines (SVM) were selected in most of the cases [35, 37, 38, 40, 43–45]. The outcomes derived from these studies presented a classification accuracy (Acc) ranging between 73% and 95%, being the frontal and parietal brain regions the most relevant for the detection of emotional states with these regularity-based entropies.

On the other hand, the predictability-based entropy indices have been the most applied for the assessment of different emotional states from EEG recordings. The main studies that have used these entropy metrics for that purpose are included in **Tables 2** and **3**. It can be observed that only a few works studied these indices between 2011 and 2015 [49–52, 63]. Nevertheless, the interest in these predictability measures has notably increased since 2017 until nowadays. More precisely, DEn is the predictability-based metric that has gained a considerable visibility since 2018, thus **Table 3** only includes studies based on the application of DEn. The rest of predictability-based entropy metrics, i.e. ShEn, REn, SpEn, and permutation indices, are contained in **Table 2**. It is interesting to note that the majority of these works in **Table 2** analyzed the EEG signals contained in the DEAP database, whereas only a few tested those metrics with different experiments. As for the regularity-based indices, the number of emotions identified in works in **Table 2**

Ref. (Year)	Experimental design	Features	Classifier	Results
[35] (2011)	2 emotions, 15 subjects, 5 EEG channels, images	ApEn + others	SVM[1]	Acc = 73.25%
[36] (2014)	Depression, 60 subjects, 24 EEG channels, eyes open/closed, no stimuli	ApEn + others	—	Higher irregularity for healthy than for depressed
[37] (2014)	4 emotions, DEAP[2]	SampEn	SVM	Acc = 80.43%
[38] (2016)	4 emotions, DEAP	SampEn + others	SVM	Acc = 94.98% (two classes) and 93.20% (four classes)
[39] (2016)	2 emotions, DEAP	SampEn, QSampEn + others	DT[3]	Acc = 75.29%
[40] (2017)	3 emotions, 44 subjects, 31 EEG channels, images	ApEn + others	SVM	Acc = 75.5%
[41] (2018)	Valence and arousal, DEAP	SampEn + others	MLP[4], DST[5]	Acc = 87.43% (arousal) and 88.74% (valence)
[42] (2018)	4 emotions, DEAP	SampEn + others	PSAE[6]	Acc = 93.6%
[43] (2018)	4 emotions, 10 subjects, 14 EEG channels, film clips	ApEn + others	SVM, DBN[7]	Acc = 87.32%
[44] (2019)	4 emotions, 8 subjects, 12 EEG channels, music	ApEn, SampEn + others	SVM, C4.5, LDA[8]	Acc = 84.91% (valence) and 89.65% (arousal)
[45] (2020)	3 emotions, SEED[9]	DySampEn[10]	SVM	Acc = 84.67%
[46] (2021)	2 emotions, DEAP	CSampEn[11]	—	More coordination in parietal and occipital

[1]*SVM: Support vector machine.*
[2]*DEAP: Database for emotion analysis using physiological signals.*
[3]*DT: Decision tree.*
[4]*MLP: Multi-layer perceptron.*
[5]*DST: Dempster-Shafer theory.*
[6]*PSAE: Parallel stacked autoencoders.*
[7]*DBN: Deep belief networks.*
[8]*LDA: Linear discriminant analysis.*
[9]*SEED: SJTU emotion EEG database.*
[10]*DySampEn: Dynamic SampEn.*
[11]*CSampEn: Cross-sample entropy.*

Table 1.
Studies of emotions recognition from EEG recordings with regularity-based entropy indices.

ranged from 2 to 4, and only one study recognized 5 emotional states [49]. In terms of the classifiers implemented, SVM approaches were preferred over other models in the majority of the studies. The results obtained presented inconsistent Acc values, ranging from 65–99%, being the frontal and parietal/occipital lobes the most relevant in emotional processes. With respect to the studies in **Table 3**, it can be noticed that the majority of them followed the same experimental procedure, but in this case, another public dataset different from DEAP was selected. Indeed, recordings from the SJTU Emotion EEG Dataset (SEED) were chosen and assessed for the detection of three emotional states, namely positive, neutral and negative [63]. This database contained EEG recordings with 62 channels from 15 subjects during the visualization of film clips with emotional content [63]. The selection of the

Ref. (Year)	Experimental design	Features	Classifier	Results
[49] (2011)	5 emotions, 20 subjects, 24 and 62 EEG channels, videos	ShEn, SP[1]	LDA, KNN[2]	Entropy better than linear. Acc = 83.04% with 62 channels
[50] (2013)	Stress, 13 subjects, 3 EEG channels, eyes closed, no stimuli	REn + others	ANOVA[3]	Lower complexity in stress than in calmness
[51] (2015)	4 emotions, 8 subjects, 20 EEG channels, audiovisual stimuli	ShEn, REn	MC-LSSVM[4]	Acc = 84.79%
[52] (2015)	4 emotions, 25 subjects, 3 EEG channels, music	SpEn	SVM, KNN, CFNN[5]	Acc = 93.66% (valence) and 93.29% (arousal)
[53] (2017)	2 emotions, DEAP	PerEn, AAPE, QSampEn	SVM	Acc = 81.31%
[54] (2017)	Valence and arousal, DEAP	SpEn, ShEn	LSSVM[6], D-RFE[7]	Acc = 78.96% (arousal) and 71.43% (valence)
[55] (2017)	4 emotions, DEAP	SpEn, ShEn + others	Three-stage decision method	Acc = 86.67%
[56] (2018)	Arousal and valence, DEAP	SpEn, spectral and statistics	SVM, KNN, NB[8]	Spectral and statistics better than SpEn
[57] (2018)	2, 3, 4 and 5 emotions, DEAP	REn + others	SVM	Acc = 73.8–86.2%
[58] (2018)	Depression, 213 subjects, 3 EEG channels, sounds	ShEn, SpEn + others	KNN	Acc = 79.27%
[59] (2019)	4 emotions, DEAP	ShEn, SpEn + others	LSSVM	Acc = 65.13%
[60] (2019)	4 emotions, DEAP	ShEn, PerEn + others	SVM	Best results with PerEn
[61] (2020)	2 emotions, DEAP	CEn[9], QSampEn	SVM	Acc = 80.31%
[62] (2020)	4 emotions, DEAP	SpEn, ShEn + others	SVM, NB, ANN[10]	Acc = 98.7% with ANN
[48] (2021)	4 emotions, DEAP	AAPE, PerMin + others	SVM	Acc = 96.39%

[1]*SP: Spectral power.*
[2]*KNN: K-nearest neightbor.*
[3]*ANOVA: Analysis of variance.*
[4]*MC-LSSVM: Multiclass least-square support vector machine.*
[5]*CFNN: Cascade-forward neural network.*
[6]*LSSVM: Least-square support vector machine.*
[7]*D-RFE: Dynamical recursive feature elimination.*
[8]*NB: Naive Bayes.*
[9]*CEn: Conditional entropy.*
[10]*ANN: Artificial neural network.*

Table 2.
Studies of emotions recognition from EEG recordings with predictability-based entropy indices (ShEn, SpEn, REn, PerEn, AAPE.

classification approaches was quite inconsistent across the different studies. However, it can be observed that deep learning approaches like convolutional neural networks (CNN) have been progressively introduced in the literature for their application in emotion recognition researches. The accuracy results reported in

Ref. (Year)	Experimental design	Features	Classifier	Results
[63] (2015)	3 emotions, SEED	DEn, SP, statistics	DBN	DEn better than SP and statistics. Acc = 85%
[64] (2018)	4 emotions, two experiments: DEAP and SEED	DEn + others	GELM[1]	Acc = 69.67% (DEAP) and 91.07% (SEED)
[65] (2018)	3 emotions, SEED	DEn	HCNN[2]	Best results in β and γ
[66] (2018)	3 emotions, 14 subjects, 64 EEG channels, film clips	DEn + others	GRSLR[3]	Acc = 80.27%
[67] (2018)	3 emotions, two experiments: DEAP and SEED	DEn	—	Best results with SEED database
[68] (2018)	3 emotions, SEED	DEn, SP, statistics	DGCNN[4]	DEn better than SP and statistics. Acc = 90%
[69] (2019)	3 emotions, SEED	DEn	LDA	Acc = 68%
[70] (2019)	2 emotions in patients with disorder of conciousness, 18 subjects, 32 EEG channels, videos	DEn	SVM	Acc = 91.5%
[71] (2019)	3 emotions, SEED	DEn	STNN[5]	Acc = 84.16%
[72] (2019)	3 emotions, SEED	DEn	LR[6]	Acc = 86%
[73] (2020)	3 emotions, SEED	DEn	MTL[7]	Acc = 88.92%
[74] (2020)	3 emotions, SEED	DEn	CNN	Acc = 90.63%
[75] (2020)	3 emotions, SEED	DEn	CNN	Acc = 90.41%
[76] (2020)	High-low valence and arousal, DEAP	DEn + others	LORSAL[8]	Acc = 77.17%
[77] (2020)	3 emotions, SEED	DEn + others	CNN	Acc = 99.7%
[78] (2020)	3 emotions, SEED	DEn + others	SRU[9]	Acc = 83.13%
[79] (2021)	Valence and arousal, DEAP	DEn	CNN	Acc = 90.45% (valence) and 90.6% (arousal)

[1]*GELM: Graph-regularized extreme learning machine.*
[2]*HCNN: Hierarchical convolutional neural network.*
[3]*GRSLR: Graph regularized sparse linear regression.*
[4]*DGCNN: Dynamical graph convolutional neural network.*
[5]*STNN: Spatial–temporal neural network.*
[6]*LR: Linear regression.*
[7]*MTL: Multisource transfer learning.*
[8]*LORSAL: Logistic regression via variable splitting and augmented Lagragian.*
[9]*SRU: Simple recurrent unit network.*

Table 3.
Studies of emotions recognition from EEG recordings with predictability-based entropy indices (DEn).

these works were between 68% and 99%. Furthermore, some of them demonstrated that DEn was more suitable than some linear metrics for the identification of emotions [63, 68].

Finally, **Table 4** shows the main works focused on the application of multiscale and multilag entropy approaches for detecting emotions with EEG recordings. As can be observed, the study of these indices has emerged in the last few years, especially since 2019 until nowadays. In these works, the number of emotional states studied ranged from 2 to 5, which is in line with the rest of entropy metrics

Ref. (Year)	Experimental design	Features	Classifier	Results
[80] (2016)	5 emotions, 30 subjects, 6 EEG channels, video clips	MMSampEn[1]	—	Higher irregularity for higher arousal levels
[81] (2019)	2 emotions, DEAP	CMQSampEn[2], CMAAPE[3]	SVM, DT	Acc = 86.35%
[82] (2019)	4 emotions, DEAP	WEn	SVM, FCM[4]	Acc = 73.32%
[83] (2019)	2 emotions, DEAP	PerMin, DPerEn[5]	KNN	Acc = 92.32%
[84] (2020)	3 emotions, 10 subjects, 14 EEG channels, video clips	WEn	ANN	Acc = 98%
[85] (2021)	3 emotions, SEED	MSpEn[6] + others	ARF[7]	Acc = 94.4%
[86] (2021)	Enjoyment, 28 subjects, 8 EEG channels, art pieces	MSampEn	SVM, RVM[8]	Acc = 91.18%

[1]*MMSampEn: Multivariate-multiscale SampEn.*
[2]*CMQSampEn: Composite multiscale QSampEn.*
[3]*CMAAPE: Composite multiscale AAPE.*
[4]*FCM: Fuzzy cognitive map.*
[5]*DPerEn: Delayed PerEn.*
[6]*MSpEn: Multiscale SpEn.*
[7]*ARF: Autoencoder based random forest.*
[8]*RVM: Relevance vector machine.*

Table 4.
Studies of emotions recognition from EEG recordings with multiscale and multilag entropy indices.

evaluated. Furthermore, the signals from the DEAP database were also chosen by some of the studies included in the table. It can be noticed that the selection of the classification models was slightly inconsistent among the different studies, although SVM and deep learning approaches were the most selected. As in the previous cases, the final outcomes obtained presented accuracy values with a high variability, ranging from 73–98%.

4. General findings

The application of entropy metrics for the recognition of emotions from EEG signals has received increasing attention in the last years, reporting valuable insights about the brain's performance under different emotional conditions. However, the high variability of the results obtained could be justified by various aspects. On the one hand, the experimentation is different for each study, since there are no gold standards of experimental procedures. In this sense, the number of participants and their gender, age, or cultural characteristics, are very incosistent among studies, thus the results may not be representative of the whole population. In addition, the type of stimulus used for emotions elicitation (images, sounds, videos, etc.) is also inconsistent, since there is no consensus about which is the optimum option for triggering a strong emotional response [87]. The duration of the stimulus is another unclear point, thus different criteria are followed by each research group. Finally, although the locations of EEG electrodes are standardized, the number of EEG channels recorded is different in each experiment, ranging from 3 to 64. Moreover, some works assessed the signals corresponding to only one brain area, thus discarding the information that could be reported by the rest of regions.

All those experimental differences could bias the possibility of obtaining universal results that could be generalized to the whole population. As a consequence of all those discrepancies between experimental procedures, the studies presented in this manuscript should be carefully interpreted and compared. In addition, as not all the publications give a thorough description of their methodology, their experiments could not be reproduced by other research groups. Therefore, the assessment of signals extracted from publicly available databases, like DEAP or SEED, could eliminate this limitation, since the experimental procedure would be the same for different authors. In this sense, the reproducibility and comparability of the results obtained would be guaranteed, and the differences in the outcomes would directly appear due to the diversity of analysis methods and classification approaches.

The variability of the results could also be a consequence of the intrinsic differences of the entropy metrics evaluated. Indeed, regularity-based, predictability-based, and multiscale/multilag approaches evaluate the complexity of time series from different perspectives. Therefore, the application of either one or other type of entropy index on the same problem could report completely divergent outcomes. Nevertheless, instead of considering these inequalities as contradictory, it should be regarded as a sign of complementarity between the different entropy metrics. For instance, some characteristics of a nonlinear signal could be properly assessed with regularity-based entropies, and other dynamics would be better described by predictability and symbolic entropy measures. Consequently, the selection of either one or other type of entropy index should be done taking into account the information that is wanted to be extracted from a nonstationary time series, also considering that the combination of different entropies would report a more complete description of the nonlinear processes.

The promising outcomes presented in these studies make the entropy metrics a useful tool for the recognition of emotions from EEG recordings. However, the majority of the works are mainly focused on obtaining great classification accuracy values, for which advanced classification models with hundreds of input features are implemented. Despite providing notable numerical results in many cases, the combination of such a large amount of data in complex classification schemes derives in a total loss of clinical interpretation of the results. In this regard, information about which are the most relevant brain regions, or which EEG channels do a higher contribution to the classification model, cannot be obtained. Thus, it becomes impossible to make a thorough analysis of the brain's behavior under the emotional states detected. As a result, it would be interesting to modify some methodological aspects in this kind of studies in order to ensure the clinical interpretation of the results and reveal new insights about mental processes under emotional conditions.

5. Conclusions

Given the nonlinear and nonstationary nature of the brain, entropy indices are suitable tools for a complete description of the brain dynamics in different scenarios, including the recognition of emotional states. This chapter summarizes the main recent contributions to the research field of emotions detection through the application of entropy indices for the analysis of EEG recordings. In this sense, regularity-based, predictability-based, and multiscale/multilag entropy approaches have demonstrated their capability to discern between different emotional states and discover new insights about the brain dynamics in emotional processes. Taking into account the valuable results obtained in the studies presented in this chapter, entropy metrics could become one of the first options to be considered in systems for automatic emotions identification from EEG signals.

Acknowledgements

This work was partially supported by Spanish Ministerio de Ciencia, Innovación y Universidades, Agencia Estatal de Investigación (AEI) /European Regional Development Fund (FEDER, UE) under EQC2019-006063-P, PID2020-115220RB-C21, and 2018/11744 grants, and by Biomedical Research Networking Centre in Mental Health (CIBERSAM) of the Instituto de Salud Carlos III. Beatriz García-Martínez holds FPU16/03740 scholarship from Spanish Ministerio de Educación y Formación Profesional.

Conflict of interest

The authors declare no conflict of interest.

Author details

Beatriz García-Martínez[1,2]*, Antonio Fernández-Caballero[1,2,3] and Arturo Martínez-Rodrigo[4,5]

1 Departamento de Sistemas Informáticos, Escuela Técnica Superior de Ingenieros Industriales, Universidad de Castilla-La Mancha, Albacete, Spain

2 Instituto de Investigación en Informática de Albacete, Universidad de Castilla-La Mancha, Albacete, Spain

3 CIBERSAM (Biomedical Research Networking Centre in Mental Health), Madrid, Spain

4 Facultad de Comunicación, Research Group in Electronic, Biomedical and Telecommunication Engineering, Universidad de Castilla-La Mancha, Cuenca, Spain

5 Instituto de Tecnologías Audiovisuales de Castilla-La Mancha, Universidad de Castilla-La Mancha, Cuenca, Spain

*Address all correspondence to: beatriz.gmartinez@uclm.es

References

[1] Susi Ferrarello. *Human emotions and the origins of bioethics.* Routledge, 2021.

[2] Paul Ekman. An argument for basic emotions. Cognition and Emotion, 6(3–4): 169–200, 1992.

[3] M Schröder and R Cowie. Toward emotion-sensitive multimodal interfaces: The challenge of the European Network of Excellence HUMAINE, 2005.

[4] James A Russell. A circumplex model of affect. Journal of Personality and Social Psychology, 39(6):1161–1178, 1980.

[5] Soujanya Poria, Erik Cambria, Rajiv Bajpai, and Amir Hussain. A review of affective computing: From unimodal analysis to multimodal fusion. Information Fusion, 37:98–125, 2017.

[6] Jing Han, Zixing Zhang, and Bjorn Schuller. Adversarial training in affective computing and sentiment analysis: Recent advances and perspectives. IEEE Computational Intelligence Magazine, 14 (2):68–81, 2019.

[7] Soraia M Alarcao and Manuel J Fonseca. Emotions recognition using EEG signals: A survey. IEEE Transactions on Affective Computing, 10 (3):374–393, 2017.

[8] Maria Egger, Matthias Ley, and Sten Hanke. Emotion recognition from physiological signal analysis: A review. Electronic Notes in Theoretical Computer Science, 343:35–55, 2019.

[9] Andrius Dzedzickis, Artūras Kaklauskas, and Vytautas Bucinskas. Human emotion recognition: Review of sensors and methods. *Sensors*, 20(3): 592, 2020.

[10] Min-Ki Kim, Miyoung Kim, Eunmi Oh, and Sung-Phil Kim. A review on the computational methods for emotional state estimation from the human EEG. Computational and Mathematical Methods in Medicine, 2013: 573734, 2013.

[11] Yuzhen Cao, Lihui Cai, Jiang Wang, Ruofan Wang, Haitao Yu, Yibin Cao, and Jing Liu. Characterization of complexity in the electroencephalograph activity of Alzheimer's disease based on fuzzy entropy. *Chaos,* 25(8):083116, 2015.

[12] Mona Farokhzadi, Gholam-Ali Hossein-Zadeh, and Hamid Soltanian-Zadeh. Nonlinear effective connectivity measure based on adaptive neuro fuzzy inference system and Granger causality. Neuroimage, 181: 382–394, 2018.

[13] Germán Rodríguez-Bermúdez and Pedro J Garcia-Laencina. Analysis of EEG signals using nonlinear dynamics and chaos: A review. *Applied Mathematics & Information Sciences*, 9 (5): 2309, 2015.

[14] B. García-Martínez, A. Martinez-Rodrigo, R. Alcaraz, and A. Fernández-Caballero. A review on nonlinear methods using electroencephalographic recordings for emotion recognition. *IEEE Transactions on Affective Computing*, page 1, 2019.

[15] Berthold Bein. Entropy. Best Practice and Research Clinical Anaesthesiology, 20:101–109, 2006.

[16] Claude E Shannon. A mathematical theory of communication. *The Bell System Technical Journal,* 27:623–656, 1948.

[17] Peng Li, Chandan Karmakar, John Yearwood, Svetha Venkatesh, Marimuthu Palaniswami, and Changchun Liu. Detection of epileptic seizure based on entropy analysis of short-term EEG. *PloS one,* 13(3): e0193691, 2018.

[18] Hamed Azami, Daniel Abásolo, Samantha Simons, and Javier Escudero. Univariate and multivariate generalized multiscale entropy to characterise EEG signals in Alzheimer's disease. *Entropy,* 19(1):31, 2017.

[19] Jiannan Kang, Huimin Chen, Xin Li, and Xiaoli Li. EEG entropy analysis in autistic children. Journal of Clinical Neuroscience, 62:199–206, 2019.

[20] Reza Shalbaf, Colleen Brenner, Christopher Pang, Daniel M Blumberger, Jonathan Downar, Zafiris J Daskalakis, Joseph Tham, Raymond W Lam, Faranak Farzan, and Fidel Vila-Rodriguez. Non-linear entropy analysis in EEG to predict treatment response to repetitive transcranial magnetic stimulation in depression. *Frontiers in* Pharmacology, 9:1188, 2018.

[21] Holger Kantz and Thomas Schreiber. *Nonlinear time series analysis.* Cambridge University Press, 2003.

[22] S M Pincus. Approximate entropy as a measure of system complexity. *Proceedings of the National Academy of Sciences of the United States of America*, 88(6):2297–2301, 1991.

[23] JS Richman and JR Moorman. Physiological time-series analysis using approximate entropy and sample entropy. American Journal of Physiology. Heart and Circulatory Physiology, 278(6): H2039–H2049, 2000.

[24] Douglas E Lake and J Randall Moorman. Accurate estimation of entropy in very short physiological time series: The problem of atrial fibrillation detection in implanted ventricular devices. *American Journal of Physiology. Heart and Circulatory Physiology,* 300(1): H319–25, 2011.

[25] A Paraschiv-Ionescu, E Buchser, B Rutschmann, and K Aminian. Nonlinear analysis of human physical activity patterns in health and disease. *Physical Review E, Statistical, Nonlinear, and Soft Matter Physics*, 77(2 Pt 1): 021913, 2008.

[26] Alfréd Rényi. On measures of entropy and information. Technical report, Hungarian Academy of Sciences, Budapest, Hungary, 1961.

[27] L. Shi, Y. Jiao, and B. Lu. Differential entropy feature for EEG-based vigilance estimation. In *Proc. 35th Annual Int. Conf. of the IEEE Engineering in Medicine and Biology Society (EMBC)*, pages 6627–6630, 2013.

[28] Christoph Bandt and Bernd Pompe. Permutation entropy: A natural complexity measure for time series. *Physical Review Letters,* 88(17): 174102, 2002.

[29] Hamed Azami and Javier Escudero. Amplitude-aware permutation entropy: Illustration in spike detection and signal segmentation. Computer Methods and Programs in Biomedicine, 128:40–51, 2016.

[30] J. Fell and J. Roschke. Discrimination of sleep stages: A comparison between spectral and nonlinear EEG measures. Electroencephalography and Clinical Neurophysiology, 98(5):401–410, 1996.

[31] W. W. Burggren. Assessing physiological complexity. The Journal of Experimental Biology, 208:3221–3232, 2005.

[32] Ingrid Daubechies. The wavelet transform, time-frequency localization and signal analysis. IEEE Transactions on Information Theory, 36(5):961–1005, 1990.

[33] Farhad Kaffashi, Ryan Foglyano, Christopher G Wilson, and Kenneth A Loparo. The effect of time delay on approximate & sample entropy calculations. *Physica D: Nonlinear Phenomena,* 237(23):3069–3074, 2008.

[34] Luciano Zunino, Felipe Olivares, and Osvaldo A. Rosso. Permutation min-entropy: An improved quantifier for unveiling subtle temporal correlations. Europhysics Letters, 109: 10005, 2015.

[35] Seyyed Abed Hosseini and Mohammad Bagher Naghibi-Sistani. Emotion recognition method using entropy analysis of EEG signals. *International Journal of Image, Graphics and Signal Processing,* 3(5):30, 2011.

[36] Subha D Puthankattil and Paul K Joseph. Analysis of EEG signals using wavelet entropy and approximate entropy: A case study on depression patients. *International Journal of Medical, Health, Pharmaceutical and Biomedical Engineering,* 8(7):420–424, 2014.

[37] Xiang Jie, Rui Cao, and Li Li. Emotion recognition based on the sample entropy of EEG. Bio-Medical Materials and Engineering, 24(1): 1185–1192, 2014.

[38] Yong Zhang, Xiaomin Ji, and Suhua Zhang. An approach to EEG-based emotion recognition using combined feature extraction method. Neuroscience Letters, 633:152–157, 2016.

[39] Beatriz García-Martínez, Arturo Martínez-Rodrigo, Roberto Zangróniz Cantabrana, Jose Pastor García, and Raúl Alcaraz. Application of entropy-based metrics to identify emotional distress from electroencephalographic recordings. *Entropy*, 18(6):221, 2016.

[40] Wen-Lin Chu, Min-Wei Huang, Bo-Lin Jian, and Kuo-Sheng Cheng. Analysis of EEG entropy during visual evocation of emotion in schizophrenia. *Annals of General Psychiatry*, 16(1):34, 2017.

[41] Morteza Zangeneh Soroush, Keivan Maghooli, Seyed Kamaledin Setarehdan, and Ali Motie Nasrabadi. A novel method of EEG-based emotion recognition using nonlinear features variability and Dempster–Shafer theory. *Biomedical Engineering: Applications, Basis and Communications*, 30(04): 1850026, 2018.

[42] Sara Bagherzadeh, K Maghooli, J Farhadi, and M Zangeneh Soroush. Emotion recognition from physiological signals using parallel stacked autoencoders. *Neurophysiology*, 50(6): 428–435, 2018.

[43] Tian Chen, Sihang Ju, Xiaohui Yuan, Mohamed Elhoseny, Fuji Ren, Mingyan Fan, and Zhangang Chen. Emotion recognition using empirical mode decomposition and approximation entropy. Computers & Electrical Engineering, 72:383–392, 2018.

[44] Yimin Hou and Shuaiqi Chen. Distinguishing different emotions evoked by music via electroencephalographic signals. Computational Intelligence and Neuroscience, 2019:1–18, 2019.

[45] Yun Lu, Mingjiang Wang, Wanqing Wu, Yufei Han, Qiquan Zhang, and Shixiong Chen. Dynamic entropy-based pattern learning to identify emotions from EEG signals across individuals. Measurement, 150: 107003, 2020.

[46] Beatriz García-Martínez, Antonio Fernández-Caballero, Raúl Alcaraz, and Arturo Martínez-Rodrigo. Cross-sample entropy for the study of coordinated brain activity in calm and distress conditions with electroencephalographic recordings. *Neural Computing and Applications,* 2021.

[47] Sander Koelstra, Christian Mühl, Mohammad Soleymani, Jong-Seok Lee, Ashkan Yazdani, Touradj Ebrahimi, Thierry Pun, Anton Nijholt, and Ioannis Patras. DEAP: A database for emotion analysis using physiological signals.

IEEE Transactions on Affective Computing, 3(1):18–31, 2012.

[48] Beatriz García-Martínez, Antonio Fernández-Caballero, Luciano Zunino, and Arturo Martínez-Rodrigo. Recognition of emotional states from EEG signals with nonlinear regularity- and predictability-based entropy metrics. Cognitive Computation, 13: 403–417, 2021.

[49] Murugappan Murugappan, Ramachandran Nagarajan, and Sazali Yaacob. Combining spatial filtering and wavelet transform for classifying human emotions using EEG signals. Journal of Medical and Biological Engineering, 31 (1):45–51, 2011.

[50] Hong Peng, Bin Hu, Fang Zheng, Dangping Fan, Wen Zhao, Xuebin Chen, Yongxia Yang, and Qingcui Cai. A method of identifying chronic stress by EEG. Personal and Ubiquitous Computing, 17(7):1341–1347, 2013.

[51] Varun Bajaj and Ram Bilas Pachori. Detection of human emotions using features based on the multiwavelet transform of EEG signals. In *Brain-Computer Interfaces*. Springer, 2015.

[52] Mohsen Naji, Mohammd Firoozabadi, and Parviz Azadfallah. Emotion classification during music listening from forehead biosignals. *Signal, Image and Video Processing*, 9(6): 1365, 2015.

[53] Beatriz García-Martínez, Arturo Martínez-Rodrigo, Roberto Zangróniz, José Manuel Pastor, and Raúl Alcaraz. Symbolic analysis of brain dynamics detects negative stress. *Entropy*, 19(5): 196, 2017.

[54] Zhong Yin, Lei Liu, Li Liu, Jianhua Zhang, and Yagang Wang. Dynamical recursive feature elimination technique for neurophysiological signal-based emotion recognition. *Cognition, Technology & Work*, 19(4):667, 2017.

[55] Jing Chen, Bin Hu, Yue Wang, Philip Moore, Yongqiang Dai, Lei Feng, and Zhijie Ding. Subject-independent emotion recognition based on physiological signals: A three-stage decision method. *BMC Medical Informatics and Decision Making*, 17(3):45, 2017.

[56] L. Piho and T. Tjahjadi. A mutual information based adaptive windowing of informative EEG for emotion recognition. IEEE Transactions on Affective Computing, 11(4):722–735, 2018.

[57] Rami Alazrai, Rasha Homoud, Hisham Alwanni, and Mohammad Daoud. EEG-based emotion recognition using quadratic time-frequency distribution. Sensors, 18:2739, 2018.

[58] Hanshu Cai, Jiashuo Han, Yunfei Chen, Xiaocong Sha, Ziyang Wang, Bin Hu, Jing Yang, Lei Feng, Zhijie Ding, Yiqiang Chen, and Jürg Gutknecht. A pervasive approach to EEG-based depression detection. Complexity, 2018: 1–13, 2018.

[59] Jiahui Cai, Wei Chen, and Zhong Yin. Multiple transferable recursive feature elimination technique for emotion recognition based on EEG signals. Symmetry, 11:683, 2019.

[60] Abhishek Tiwari and Tiago H. Falk. Fusion of motif- and spectrum-related features for improved EEG-based emotion recognition. Computational Intelligence and Neuroscience, 2019:1–14, 2019.

[61] Beatriz García-Martínez, Arturo Martínez-Rodrigo, Antonio Fernández-Caballero, José Moncho-Bogani, and Raúl Alcaraz. Nonlinear predictability analysis of brain dynamics for automatic recognition of negative stress. Neural Computing and Applications, 32: 13221–13231, 2020.

[62] Mitul Kumar Ahirwal and Mangesh Ramaji Kose. Audio-visual stimulation

based emotion classification by correlated EEG channels. *Health and Technology*, 10: 7–23, 2020.

[63] W. Zheng and B. Lu. Investigating critical frequency bands and channels for EEG-based emotion recognition with deep neural networks. IEEE Transactions on Autonomous Mental Development, 7(3):162–175, 2015.

[64] W. Zheng, J. Zhu, and B. Lu. Identifying stable patterns over time for emotion recognition from EEG. IEEE Transactions on Affective Computing, 10(3):417–429, 2018.

[65] Jinpeng Li, Zhaoxiang Zhang, and Huiguang He. Hierarchical convolutional neural networks for EEG-based emotion recognition. *Cognitive Computation*, 10(2):368, 2018.

[66] Yang Li, Wenming Zheng, Zhen Cui, Yuan Zong, and Sheng Ge. EEG emotion recognition based on graph regularized sparse linear regression. Neural Processing Letters, 49:555–571, 2018.

[67] Z. Lan, O. Sourina, L. Wang, R. Scherer, and G. R. Müller-Putz. Domain adaptation techniques for EEG-based emotion recognition: A comparative study on two public datasets. IEEE Transactions on Cognitive and Developmental Systems, 11(1):85–94, 2018.

[68] T. Song, W. Zheng, P. Song, and Z. Cui. EEG emotion recognition using dynamical graph convolutional neural networks. IEEE Transactions on Affective Computing, 11(3):532–541, 2018.

[69] Dong-Wei Chen, Rui Miao, Wei-Qi Yang, Yong Liang, Hao-Heng Chen, Lan Huang, Chun-Jian Deng, and Na Han. A feature extraction method based on differential entropy and linear discriminant analysis for emotion recognition. Sensors, 19:1631, 2019.

[70] Haiyun Huang, Qiuyou Xie, Jiahui Pan, Yanbin He, Zhenfu Wen, Ronghao Yu, and Yuanqing Li. An EEG-based brain computer interface for emotion recognition and its application in patients with disorder of consciousness. IEEE Transactions on Affective Computing, pages 1–1, 2019.

[71] Y. Li, W. Zheng, L. Wang, Y. Zong, and Z. Cui. From regional to global brain: A novel hierarchical spatial-temporal neural network model for EEG emotion recognition. *IEEE Transactions on Affective Computing*, page 1, 2019.

[72] Soheil Keshmiri, Masahiro Shiomi, and Hiroshi Ishiguro. Entropy of the multi-channel EEG recordings identifies the distributed signatures of negative, neutral and positive affect in whole-brain variability. Entropy, 21: 1228, 2019.

[73] J. Li, S. Qiu, Y.-Y. Shen, C.-L. Liu, and H. He. Multisource transfer learning for cross-subject EEG emotion recognition. IEEE Transactions on Cybernetics, 50(7):3281–3293, 2020.

[74] Z. Gao, X. Wang, Y. Yang, Y. Li, K. Ma, and G. Chen. A channel-fused dense convolutional network for EEG-based emotion recognition. *IEEE Transactions on Cognitive and Developmental Systems*, page 1, 2020.

[75] Sunhee Hwang, Kibeom Hong, Guiyoung Son, and Hyeran Byun. Learning CNN features from DE features for EEG-based emotion recognition. Pattern Analysis and Applications, 23:1323–1335, 2020.

[76] Chao Pan, Cheng Shi, Honglang Mu, Jie Li, and Xinbo Gao. EEG-based emotion recognition using logistic regression with Gaussian kernel and Laplacian prior and investigation of critical frequency bands. Applied Sciences, 10:1619, 2020.

[77] Yingdong Wang, Qingfeng Wu, Chen Wang, and Qunsheng Ruan. DE-

CNN: An improved identity recognition algorithm based on the emotional electroencephalography. Computational and Mathematical Methods in Medicine, 2020:7574531, 2020.

[78] Chen Wei, Lan lan Chen, Zhen zhen Song, Xiao guang Lou, and Dong dong Li. EEG-based emotion recognition using simple recurrent units network and ensemble learning. *Biomedical Signal Processing and Control,* 58:101756, 2020.

[79] Yongqiang Yin, Xiangwei Zheng, Bin Hu, Yuang Zhang, and Xinchun Cui. EEG emotion recognition using fusion model of graph convolutional neural networks and LSTM. Applied Soft Computing Journal, 100:106954, 2021.

[80] Yelena Tonoyan, David Looney, Danilo P Mandic, and Marc M Van Hulle. Discriminating multiple emotional states from EEG using a data-adaptive, multiscale information-theoretic approach. *International Journal of Neural Systems*, 26(02):1650005, 2016.

[81] Arturo Martínez-Rodrigo, Beatriz García-Martínez, Raúl Alcaraz, Pascual González, and Antonio Fernández-Caballero. Multiscale entropy analysis for recognition of visually elicited negative stress from EEG recordings. *International Journal of Neural Systems*, 29(2):1850038, 2019.

[82] Kairui Guo, Rifai Chai, Henry Candra, Ying Guo, Rong Song, Hung Nguyen, and Steven Su. A hybrid fuzzy cognitive map/support vector machine approach for EEG-based emotion classification using compressed sensing. International Journal of Fuzzy Systems, 21:263–273, 2019.

[83] Arturo Martínez-Rodrigo, Beatriz García-Martínez, Luciano Zunino, Raúl Alcaraz, and Antonio Fernáindez-Caballero. Multi-lag analysis of symbolic entropies on EEG recordings for distress recognition. *Frontiers in Neuroinformatics*, 13(40), 2019.

[84] Qiang Gao, Chu han Wang, Zhe Wang, Xiao lin Song, En zeng Dong, and Yu Song. EEG based emotion recognition using fusion feature extraction method. *Multimedia Tools and Applications,* 79:27057–27074, 2020.

[85] A. Bhattacharyya, R. K. Tripathy, L. Garg, and R. B. Pachori. A novel multivariate-multiscale approach for computing EEG spectral and temporal complexity for human emotion recognition. IEEE Sensors Journal, 21 (3): 3579–3591, 2021.

[86] M. Fraiwan, M. Alafeef, and F. Almomani. Gauging human visual interest using multiscale entropy analysis of EEG signals. Journal of Ambient Intelligence and Humanized Computing, 12:2435–2447, 2021.

[87] Gaetano Valenza, Antonio Lanata, and Enzo Pasquale Scilingo. The role of nonlinear dynamics in affective valence and arousal recognition. *IEEE Transactions on Affective Computing*, 3 (2):237–249, 2012.

Section 3

Brain-Computer Interface Ergonomic Controllers

Chapter 8

Evaluating Steady-State Visually Evoked Potentials-Based Brain-Computer Interface System Using Wavelet Features and Various Machine Learning Methods

Ebru Sayilgan, Yilmaz Kemal Yuce and Yalcin Isler

Abstract

Steady-state visual evoked potentials (SSVEPs) have been designated to be appropriate and are in use in many areas such as clinical neuroscience, cognitive science, and engineering. SSVEPs have become popular recently, due to their advantages including high bit rate, simple system structure and short training time. To design SSVEP-based BCI system, signal processing methods appropriate to the signal structure should be applied. One of the most appropriate signal processing methods of these non-stationary signals is the Wavelet Transform. In this study, we investigated both the effect of choosing a mother wavelet function and the most successful combination of classifier algorithm, wavelet features, and frequency pairs assigned to BCI commands. SSVEP signals that were recorded at seven different stimulus frequencies (6–6.5 – 7 – 7.5 – 8.2 – 9.3 – 10 Hz) were used in this study. A total of 115 features were extracted from time, frequency, and time-frequency domains. These features were classified by a total of seven different classification processes. Classification evaluation was presented with the 5-fold cross-validation method and accuracy values. According to the results, (I) the most successful wavelet function was Haar wavelet, (II) the most successful classifier was Ensemble Learning, (III) using the feature vector consisting of energy, entropy, and variance features yielded higher accuracy than using one of these features alone, and (IV) the highest performances were obtained in the frequency pairs with "6–10", "6.5–10", "7–10", and "7.5–10" Hz.

Keywords: steady-state visually-evoked potentials (SSVEP), brain-computer interfaces (BCI), wavelet transform (WT), mother wavelet selection, pattern recognition, machine learning (ML)

1. Introduction

Electroencephalogram (EEG) signals are one of the most widely used types of biomedical signals for Brain-Computer Interfaces (BCIs), owing to their portability, high time resolution, ease of acquisition, and cost-effectiveness as compared to other brain activity monitoring techniques [1–3]. There are four typical EEG-based BCI paradigms: steady-state visual-evoked potentials (SSVEP), slow cortical potentials (SCP), the P300 component of evoked potentials, and sensory-motor rhythms (SMR) [4–6].

The SSVEP signal is a periodic response to a visual stimulator modulated at a frequency greater than 6 Hz [7] or 4 Hz [8]. The amplitude and phase characteristics of the SSVEP depend on stimulus intensity and frequency. SSVEP events can be repeatedly produced if the stimuli are provided under controlled conditions [9]. For instance, staring at a flickering light that flashes at a constant frequency stimulates the human visual pathway. The flickering frequency is radiated throughout the brain. This stimulation produces electrical signals in the brain at the base frequency of the flashing light, as well as at its harmonics [10]. Practically, there is a marked reduction in the power of the SSVEP signals from the second harmonics onwards. This has been attributed to the low signal-to-noise ratio of the SSVEP signals at high frequencies and can be accounted for the brain dynamics that act as a low pass filter [11].

The analysis of EEG signals using machine learning (ML) methods is developed to help physicians in accurate diagnosis and provides fast and valid tools in assistive applications designed for individuals. Among the various approaches available in the literature, the Wavelet Transform (WT) has proven to be an effective time-frequency analysis tool for analyzing transient signals [12, 13]. Various wavelet families are available to define and adapt signal characteristics [14]. However, choosing an appropriate mother wavelet is very important for the analysis of these signals. Research studies to date for EEG-signal classification using the wavelet technique have mostly been done using the Daubechies (Db) family. The maximum accuracy achieved in this study was 95.00% [15]. However, in this study, although the signal was suitable for Discrete Wavelet Transformation (DWT), analysis was performed using the Continuous Wavelet Transformation (CWT) method. Furthermore, in the same study, analysis was performed for a single frequency. In this chapter, a detailed analysis was performed using multiple frequencies. Also, in Ref. [16], the SSVEP signal was used for a single wavelet type (Db40), but no mother wavelet selection was made. Thus, the mother wavelet selection for SSVEP is still an unanswered question.

The research presented in this chapter is especially about selecting the most suitable wavelet function for signal analysis of SSVEP signals, detailed investigation of energy, entropy, and variance attributes, and examining the appropriate frequency(s) for SSVEP based BCI design.

There is not any, to our knowledge, in-depth study on the selection of stimulation frequencies. It was noticed that higher accuracy rates could be obtained for pattern recognition by examining the frequency selection and the differences between the frequencies. The frequency or frequencies that might result in higher higher accuracy rates and time advantages are considered to help design user-friendly BCI systems. Due to the shortcomings in the literature mentioned above, this study was considered to be conducted.

2. Materials and methods

2.1 Data recording process and users

In this study, the dataset (AVI SSVEP Dataset) containing SSVEP signals designed and recorded by Adnan Vilic was used [17]. The data set contains data that include EEG measurements of healthy individuals (three men and one woman having ages range from 27 to 32) looking at the flickering target to trigger responses of SSVEP signals at different frequencies, and the data set used for this study is publicly available. Using the standard international 10–20 system for electrode placement, the reference electrode is positioned in Fz with the signal electrode in Oz and Fpz in the ground electrode. In this experiment, individuals had been seated 60 cm away from a monitor staring at a single flashing target whose color changed rapidly from black to white. The test stimulus was a flashing box at seven different frequencies (6–6.5 - 7 - 7.5 - 8.2 - 9.3 - 10 Hz) presented on the monitor. The data set comprises of four sessions with four different participants. Each trial in a session lasts 30 seconds and participants take a short break between trials. Experiments were repeated at least three times for each frequency.

In **Figure 1**, a) the raw signal stimulated at a frequency of 10 Hz and b) the power spectrum density computed signal (with its 1st and 2nd harmonics) are shown.

2.2 Feature extraction

It is possible to define the neurophysiology of the human visual system, the neuronal activity of the visual cortex is replaced by visual stimulation, and variations of the brain response related to the features of the visual stimulus such as brightness, contrast and frequency [18]. Neurons in the visual cortex synchronize their flickering to the frequency of blinking of the visual stimulus. SSVEP signals are generated when visual stimuli are repeatedly presented, creating almost sinusoidal oscillations [19]. Applying a visual stimulus flashing at a constant frequency increases the energy of brain activities comparing to the case of applying a constant visual stimulus [7]. The strongest response occurs in the visual cortex of brain (occipital), but other areas of brain are also activated to different degrees [8, 9]. SSVEP marks can be detected even for narrow frequency bands around the visual stimulation frequency with signal processing methods that take advantage of the specific features of the signal such as timing, frequency, and rhythm [20]. For this reason, this study is designed on accepted signal processing strategies that validate the comprehensive scenarios analyzed.

2.2.1 Time-domain based feature extraction

The SSVEP time-domain features are extracted from available literature information in the original field of the EEG signal. **Table 1** describes the relevant and distinctive SSVEP time-domain features we identified. These features are based on the amplitude (e.g. average amplitude change value, root mean square, interquartile

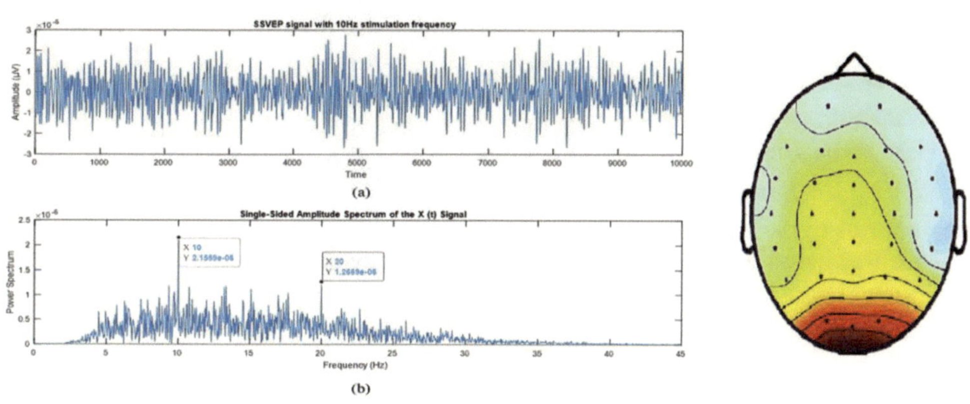

Figure 1.
a) SSVEP raw signal b) power spectrum of the 10 Hz stimulated SSVEP signal and topography.

EEG Time-domain features $\left(F_i^{(t)}\right)$			
No.	**Features**	**No.**	**Features**
1.	EEG minimum value	14.	Kurtosis of EEG signal
2.	EEG maximum value	15.	Skewness of EEG signal
3.	EEG mean value	16.	Hjorth identifiers: 1) Activity
4.	EEG standard deviation value	17.	Hjorth identifiers: 2) Mobility
5.	Integrated EEG value	18.	Hjorth identifiers: 3) Complexity
6.	Mean absolute value	19.	Signal range (max-min.)
7.	Simple square integral value	20.	Inter-quarter intervals 1st Quartile
8.	EEG variance value	21.	Inter-quarter intervals 2nd Quartile (Median)
9.	Root mean square value	22.	Inter-quarter intervals 3rd Quartile
10.	Waveform length value	23.	Zero-crossing
11.	Average amplitude change value	24.	Slope-change value
12.	Absolute difference in standard deviation	25.	Mode value of the signal

Table 1.
EEG time-domain features (EEG signal is represented by x, and $F_i^{(t)}$ *stands for the EEG features computed from x).*

ranges, etc.) and statistical changes of the EEG signal (e.g., mean, variance, skewness, and kurtosis, etc.) [20].

2.2.2 Frequency-domain based feature extraction

SSVEP signals are identified by oscillations with frequencies synchronized with the stimulus frequency [6, 21]. For this reason, many SSVEP-based BCI systems use frequency information embedded in the signal in the feature extraction process [22, 23]. Within the scope of this chapter, SSVEP frequency features were extracted from the frequency domain representation of the SSVEP signal using a Fourier transform. The relevant and distinctive SSVEP frequency characteristics we detected are based on the spectral information of SSVEP signals for each EEG rhythm, such as energy, variance and spectral entropy.

These features explain how power, variance, and irregularity (entropy) change in certain related frequency bands. In practice, this implies that these features will use their power in certain frequency bands [24].

Features based on power spectrum, energy of each frequency band,

$$F_1^{(f)} = Energy_f = \sum_{k=1}^{M} y(k)^2 \tag{1}$$

Here is the Fourier transform of the analytic signal y of a real discrete time EEG signal x.$F_1^{(f)} = Energy_f$ stands for the EEG features computed from y, and M corresponds to the maximum frequency.

Features based on variance of each EEG frequency band,

$$F_2^{(f)} = Variance_f = \frac{1}{M-1}\sum_{k=1}^{M} (y_k - \bar{y})^2 \tag{2}$$

"$\bar{y}$" in the formula stands for the average of the "y" signal.

Feature based on entropy of each EEG frequency band: Spectral entropy measures the regularity of the power spectrum of EEG signal,

$$F_3^{(f)} = Entropy_f = \frac{1}{\log(M)} \sum_{k=1}^{M} P(y(k)) \log P(y(k)) \tag{3}$$

2.2.3 Wavelet transform based feature extraction

2.2.3.1 Wavelet decomposition

SSVEP signal is non-stationary [18]. Consequently, WT has been used to examine not only spectral analysis of the signal but also the spectral behavior of the signal over time. This method is characterized by a smooth and fast oscillating function that is well localized in frequency and time [12]. WT can be applied as a specially designed dual Finite-Impulse Response (FIR) filter. The frequency responses of FIR filters separate the high frequency and low-frequency components of the input signal. The point of dividing the signal frequency is usually between 0 Hz and half the data sampling rate (Nyquist frequency). In the Multi-resolution Algorithm (MRA) of the WT, the identical wavelet coefficients are used in both low-pass (LP) and high-pass (HP) filters. The LP filter coefficients are associated with scaling parameter, which will determine the oscillatory frequency and the length of the wavelet. At the same time, the HP filter is associated with the wavelet function. The outputs of the LP filters are called the approximation *(a)* coefficients, and the outputs of the HP filters are called the detail *(d)* coefficients. In MRA of WT, any time-series signals can be entirely decomposed in terms of a and d coefficients based on decomposition level. Implementation of DWT on raw signal produces an MRA of various statistical and non-statistical parameters across time and frequency [24]. The subsets of the wavelet coefficients of the decomposition tree were selected as input vectors to the classifier. The SSVEP signals are decomposed into 9 decomposition levels, and $i = 1, 2, \ldots, 9$ for 512 Hz sampling frequency.

2.2.3.2 Parameters for feature extraction

Using different DWT functions (Haar, Db2, Sym4, Coif1, Bior3.5, Rbior2.8), SSVEP signals are subdivided into frequency bands (delta, theta, alpha, beta, gamma), and the energy, entropy and variance were calculated for each band [13, 14]. Every DWT frequency band is associated with one or two EEG rhythms. Thus, a number of features represented in the frequency bands were obtained.

Energy at each decomposition level was calculated using the following Equations [24]:

$$F_1^{(w)} = Ed_i = \sum_{j=1}^{N} |d_{ij}|^2, i = 1, 2, 3, \ldots, l \tag{4}$$

$$F_1^{(w)} = Ea_i = \sum_{j=1}^{N} |a_{ij}|^2, i = 1, 2, 3, \ldots, l \tag{5}$$

where d_{ij} and a_{ij} represent detail and approximate coefficients, respectively, formed by the wavelet level corresponding to each EEG band (delta, theta, alpha, beta, gamma).$i = 1, 2, 3, \ldots, l$ is the wavelet decomposition level from levels 1 to l. Finally, N stands for the number of detail and approximate coefficients at each decomposition level.

Another feature, the entropy at each decomposition level is calculated using the following Equation [25]:

$$F_2^{(w)} = Ent_i = -\sum_{j=1}^{N} {d_{ij}}^2 \log\left({d_{ij}}^2\right), i = 1, 2, 3, \dots, l \tag{6}$$

The variance at each decomposition level was calculated using the following Equation [24]:

$$F_3^{(w)} = Var_i = \frac{1}{N-1}\sum_{j=1}^{N} \left(d_{ij} - \mu_i\right)^2, \mu_i = \frac{1}{N}\sum_{j=1}^{N} d_{ij}, i = 1, 2, 3, \dots, l \tag{7}$$

Extracted features, which consist of different combinations, (l +1) dimensional are used as input vectors. In other words, for an 'l' level decomposition, the feature vector of any parameter can be represented as Feature = $[xd_1, xd_2, \dots, xd_l, xa_l]$, where x stands on energy, entropy, and variance.

2.3 Machine learning classification algorithms

The most important use of machine learning (ML) methods is classification [26]. After feature extraction, classification is performed to recognize an SSVEP signal and convert it to command, that is, to use it as output [27]. For the classification process, the "datasets" formed by a certain number of feature vectors, of which class it belongs, are passed through the training period required by the classification type. As a result of this training, a decision mechanism algorithm is created, which is used to assign the unknown signal to the appropriate class [28, 29].

The extracted feature vectors have been tested with seven well-known and commonly-used basic classifiers. These selected classifier algorithms are *Decision Trees, Discriminant Analysis, Logistic Regression, Naive Bayes, Support Vector Machines, k-Nearest Neighbors, and Ensemble Learner.* The classifier performances were examined to determine which combination of mother wavelet function, wavelet features, and classifier algorithm gives the highest accuracy.

2.4 Evaluation of machine learning algorithms performance

While training ML algorithm to classify SSVEP signals is an important step, it is essential to consider how the algorithm is generalized on unprecedented data (test set) [30]. We need to know if the algorithm works correctly and whether we can trust its predictions. The machine learning algorithm can only memorize the training set. Therefore, it can make reasonable predictions about future examples or examples that it has not seen before. Thus, it is one of the essential steps for BCI systems to know and apply the techniques used to evaluate how well a ML model generalizes to new, unprecedented data [31, 32]. For this goal the "k-fold cross-validation" and "confusion matrix" evaluation criteria were used to evaluate the performance of the ML algorithms used in this study.

2.4.1 k-fold cross-validation

In this method, the data set is randomly divided into k segments. Among these segments, k-1 parts are used for the training, and the remaining part is used for the testing. This process is repeated until all parts are used for testing separately. The test errors are recorded each time, and the average of the errors after the last part is reported. The performance of each classifier algorithm used is measured by carrying out this approach [30, 31]. In this study, the data set is divided into five equal parts.

2.4.2 Confusion matrix

Confusion matrix is, at first, calculated to evaluate the classifier performance. The confusion matrix is generated by comparing the responses of the classification algorithm to the test set with the actual values in the data set. In case of two-state problems, it is a table consisting of four different situations [26]. These are True Positive (TP) value, True Negative (TN) value, False Positive (FP) value, and False Negative (FN) value.

Accuracy value (ACC) is calculated as classifier performance based on these values [27]:

$$ACC = \frac{\mathrm{TP} + \mathrm{TN}}{\mathrm{TP} + \mathrm{FN} + \mathrm{FP} + \mathrm{TN}} \tag{8}$$

2.5 Experimental design and implementation details

In accordance with the objective of our study, we have designed it in a two-fold manner for time-frequency domain features. First, we measured the accuracy of each (feature, mother wavelet function) pair. As the second part, we combined the set of three features with each mother wavelet function in order to discover which mother wavelet function yields the best performance in terms of accuracy. Three important features (i.e. energy, variance, and entropy) have been extracted for EEG bands (i.e. delta, theta, alpha, beta, and gamma) using six different mother wavelet families (Haar, db, sym, coif, bior, rbio). To this purpose, algorithms were implemented using Signal Processing Toolbox and Wavelet Toolbox in Matlab 2019a. All the classifiers and performance analyses were implemented using the Classifier Learner App tool from Matlab version 2019a.

3. Results and discussion

Characterized as an increase in the amplitude of the stimulating frequency, the photic driver response results in significant baseline and harmonics [33]. Thus, it is possible to determine the stimulus frequency based on the SSVEP measurement. For this purpose, 115 feature vectors were extracted from the SSVEP signals recorded using seven different frequencies. The extracted feature vectors were run with seven basic ML algorithms. Simultaneously, the frequencies that constitute the SSVEP data set were evaluated with multiple, selected three-class, and binary classifications. Also, the effect of the increase in the difference between frequencies on the accuracy criterion was investigated, and the results are shown in detail between **Figures 2–17**, and **Tables 2–5**.

3.1 Time-domain features results

The multiple and binary classification results of 25 feature vectors extracted from SSVEP signals using time-domain properties are given below, respectively.

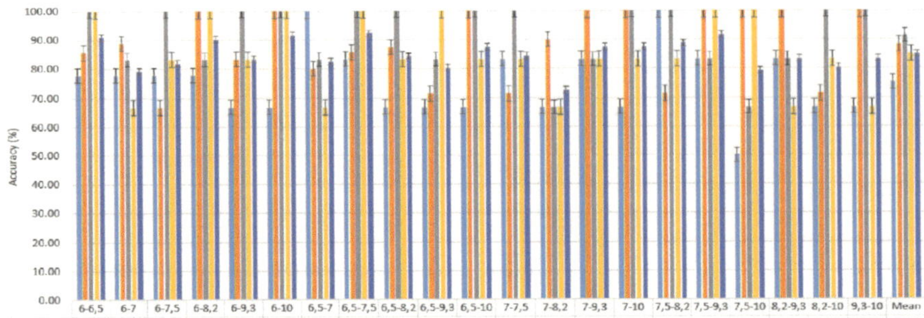

	6-6,5	6-7	6-7,5	6-8,2	6-9,3	6-10	6,5-7	6,5-7,5	6,5-8,2	6,5-9,3	6,5-10	7-7,5	7-8,2	7-9,3	7-10	7,5-8,2	7,5-9,3	7,5-10	8,2-9,3	8,2-10	9,3-10	Mean
Subject1	77.80	77.80	77.80	77.80	66.70	66.70	100.00	83.30	66.70	66.70	66.70	83.30	66.70	83.30	66.70	100.00	83.30	50.00	83.30	66.70	66.70	75.14
Subject2	85.70	88.90	66.70	100.00	83.30	100.00	80.00	85.70	87.50	71.40	100.00	71.40	90.00	100.00	100.00	71.40	100.00	100.00	100.00	71.40	100.00	88.26
Subject3	100.00	83.30	100.00	83.30	100.00	100.00	83.30	100.00	100.00	83.30	100.00	100.00	66.70	83.30	100.00	100.00	83.30	66.70	83.30	100.00	100.00	91.26
Subject4	100.00	66.70	83.30	100.00	83.30	100.00	66.70	100.00	83.30	100.00	83.30	83.30	66.70	83.30	83.30	83.30	100.00	100.00	66.70	83.30	66.70	84.91
Mean	90.88	79.18	81.95	90.28	83.33	91.68	82.50	92.25	84.38	80.35	87.50	84.50	72.53	87.48	87.50	88.68	91.65	79.18	83.33	80.35	83.35	84.89

Figure 2.
Binary classification performance of the time-domain features.

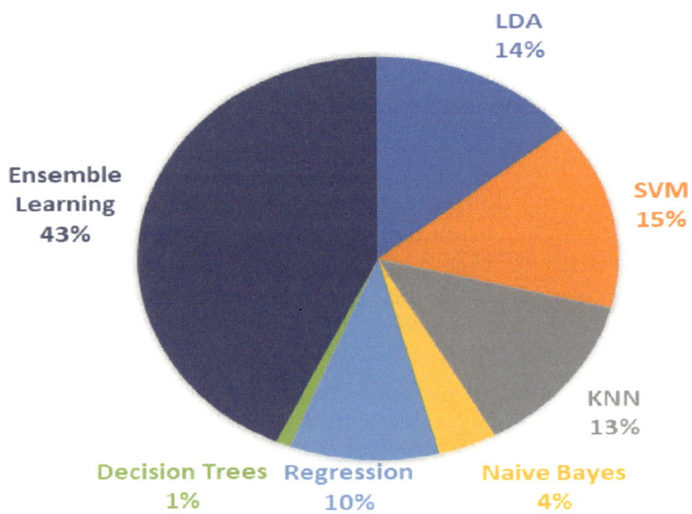

Figure 3.
Percentage of classifier where the best result is the most often obtained as a result of running the algorithms 2,520 times in total.

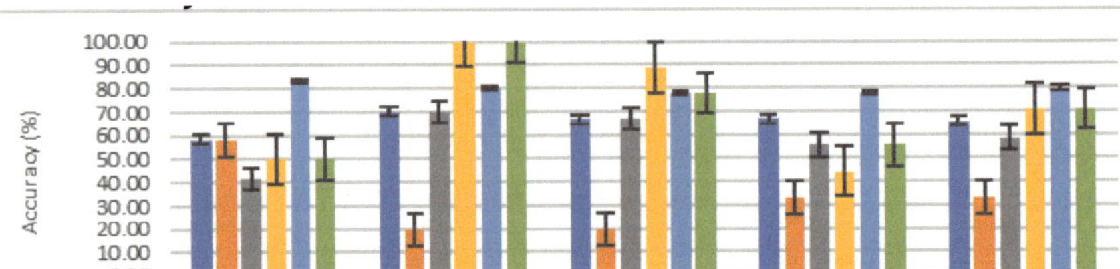

	Subject 1	Subject 2	Subject 3	Subject 4	Mean
Discriminant Analysis	58.30	70.00	66.70	66.70	65.43
Decision Trees	58.30	20.00	20.00	33.30	32.90
Naive Bayes	41.70	70.00	66.70	55.60	58.50
KNN	50.00	100.00	88.90	44.40	70.83
Ensemble Learning	83.30	80.00	77.80	77.80	79.73

Figure 4.
Results of selected 3-class classifications for frequency-domain features.

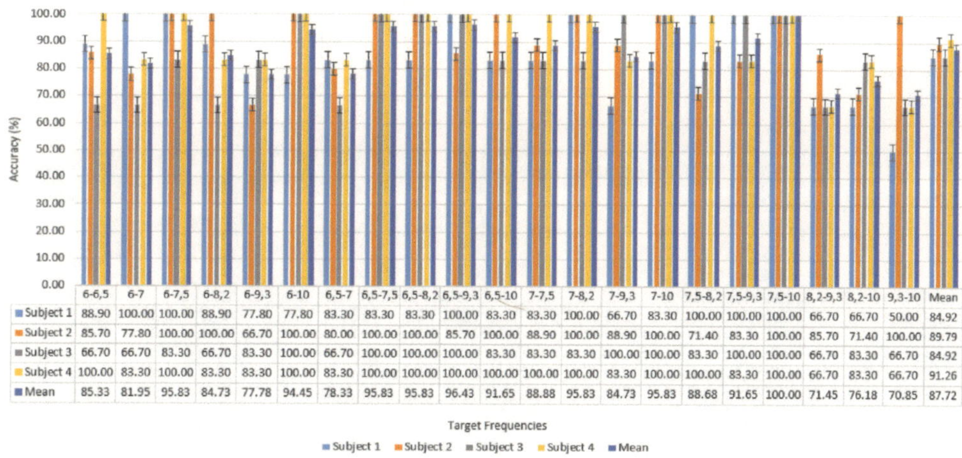

	6-6,5	6-7	6-7,5	6-8,2	6-9,3	6-10	6,5-7	6,5-7,5	6,5-8,2	6,5-9,3	6,5-10	7-7,5	7-8,2	7-9,3	7-10	7,5-8,2	7,5-9,3	7,5-10	8,2-9,3	8,2-10	9,3-10	Mean
Subject 1	88.90	100.00	100.00	88.90	77.80	77.80	83.30	83.30	83.30	100.00	83.30	83.30	100.00	66.70	83.30	100.00	100.00	100.00	66.70	66.70	50.00	84.92
Subject 2	85.70	77.80	100.00	100.00	66.70	100.00	80.00	100.00	100.00	85.70	100.00	88.90	100.00	88.90	100.00	71.40	83.30	100.00	85.70	71.40	100.00	89.79
Subject 3	66.70	66.70	83.30	66.70	83.30	100.00	66.70	100.00	100.00	100.00	83.30	83.30	83.30	100.00	100.00	83.30	100.00	100.00	66.70	83.30	66.70	84.92
Subject 4	100.00	83.30	100.00	83.30	83.30	100.00	83.30	100.00	100.00	100.00	100.00	100.00	100.00	83.30	100.00	100.00	83.30	100.00	66.70	83.30	66.70	91.26
Mean	85.33	81.95	95.83	84.73	77.78	94.45	78.33	95.83	95.83	96.43	91.65	88.88	95.83	84.73	95.83	88.68	91.65	100.00	71.45	76.18	70.85	87.72

Figure 5.
Binary classification performance of the frequency-domain features.

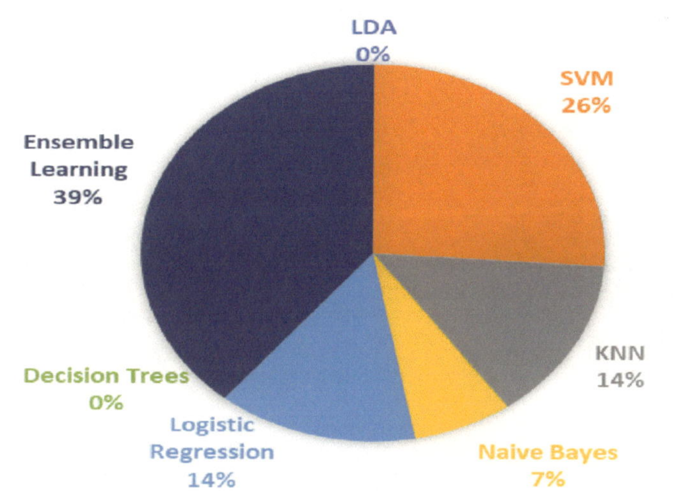

Figure 6.
Percentage of successful classifiers that give the highest accuracies from 2,520 runs in total.

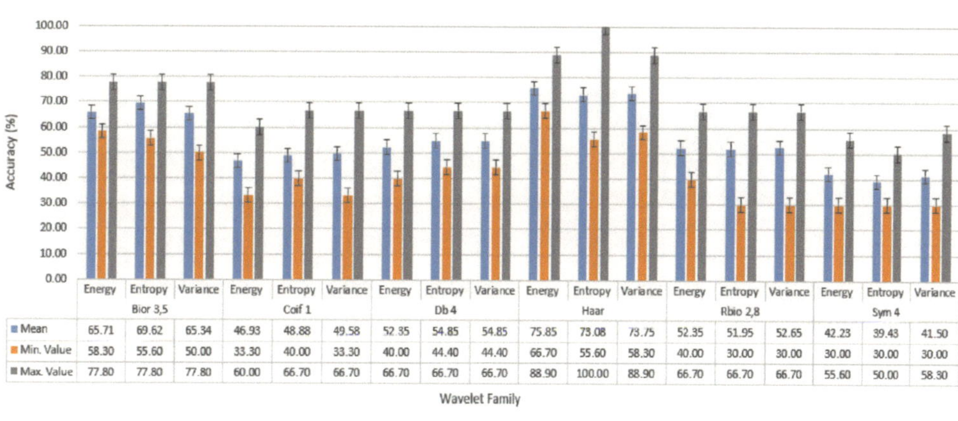

	Bior 3,5			Coif 1			Db 4			Haar			Rbio 2,8			Sym 4		
	Energy	Entropy	Variance	Energy	Entropy	Variance	Energy	Entropy	Variance	Energy	Entropy	Variance	Energy	Entropy	Variance	Energy	Entropy	Variance
Mean	65.71	69.62	65.34	46.93	48.88	49.58	52.35	54.85	54.85	75.85	73.08	73.75	52.35	51.95	52.65	42.23	39.43	41.50
Min. Value	58.30	55.60	50.00	33.30	40.00	33.30	40.00	44.40	44.40	66.70	55.60	58.30	40.00	30.00	30.00	30.00	30.00	30.00
Max. Value	77.80	77.80	77.80	60.00	66.70	66.70	66.70	66.70	66.70	88.90	100.00	88.90	66.70	66.70	66.70	55.60	50.00	58.30

Figure 7.
Classification performance of energy, entropy, and variance as separate features.

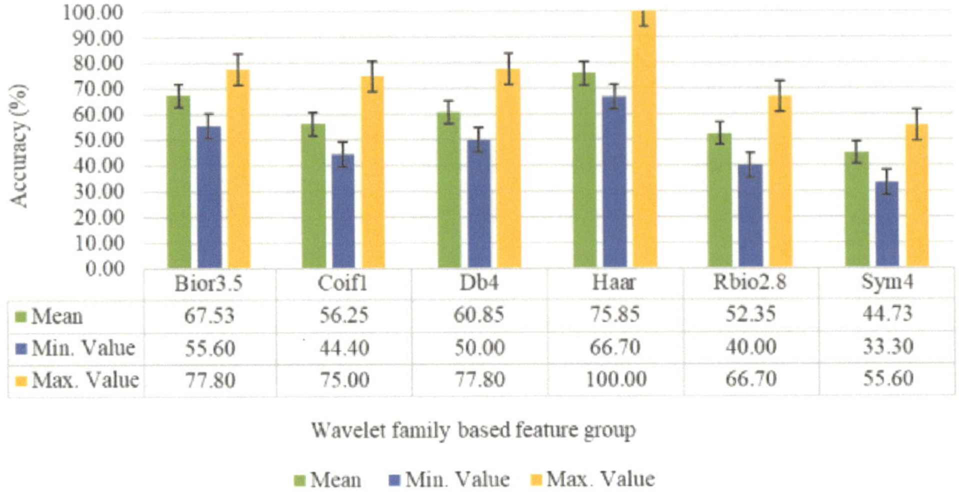

Figure 8.
Classification performance of energy, entropy, and variance together as a feature set (all features together).

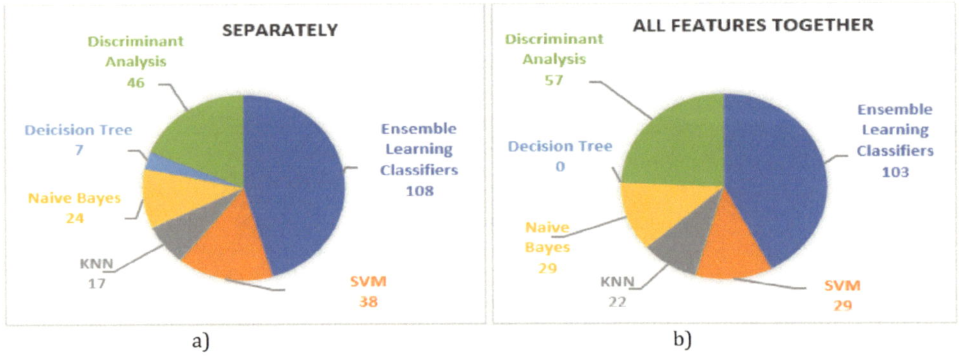

Figure 9.
Percentage of classifier where the best result is the most often obtained as a result of running the algorithms 2,520 times in total a) energy, entropy, and variance as separate features, b) energy, entropy, and variance as a feature set.

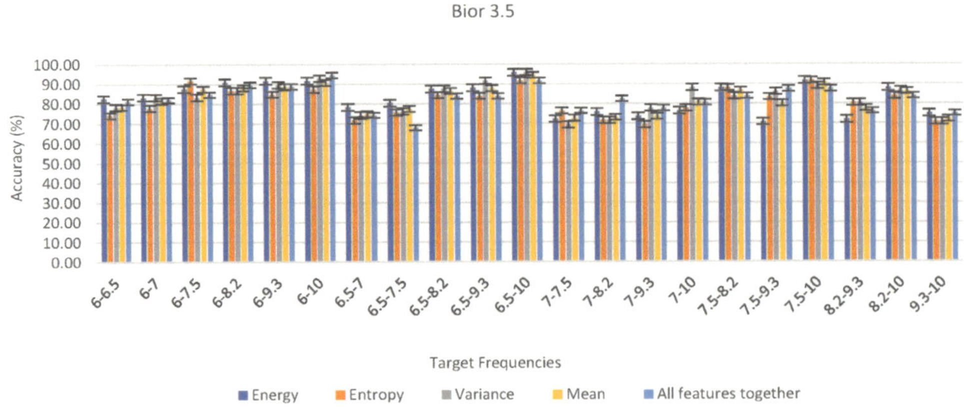

Figure 10.
Binary classification performance of the features for bior 3.5 mother wavelet function.

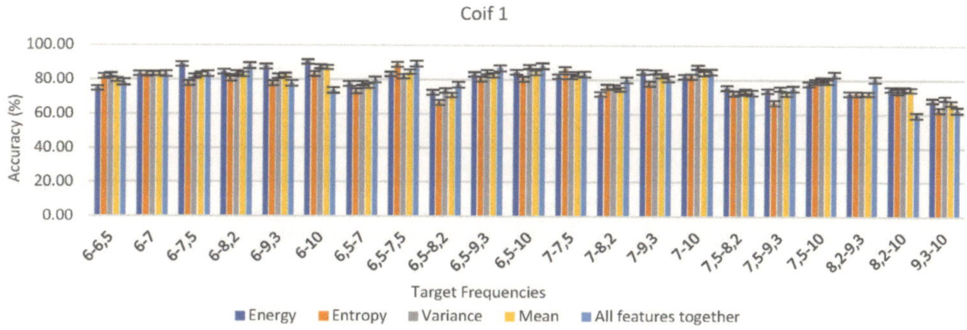

Figure 11.
Binary classification performance of the features for coif 1 mother wavelet function.

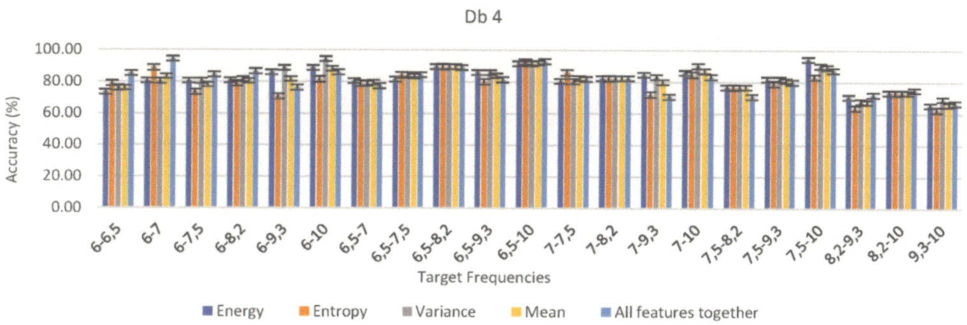

Figure 12.
Binary classification performance of the features for Db 4 mother wavelet function.

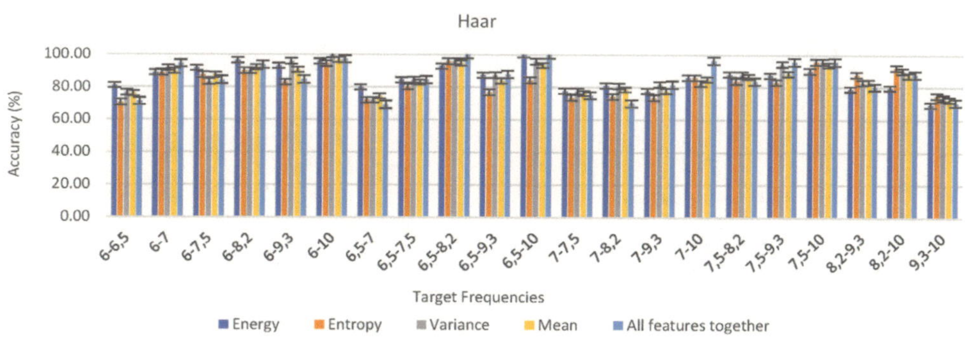

Figure 13.
Binary classification performance of the features for Haar mother wavelet function.

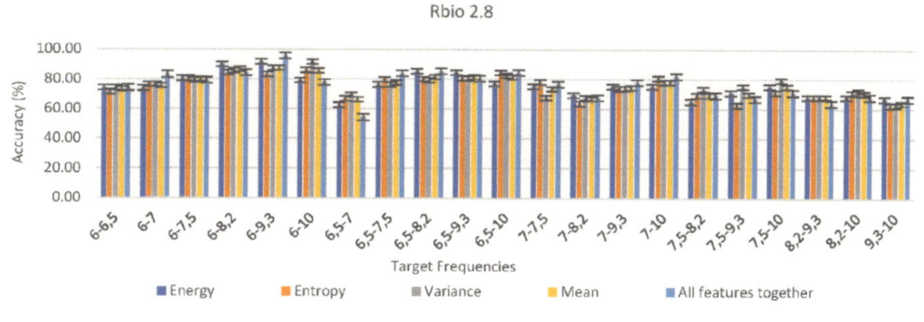

Figure 14.
Binary classification performance of the features for Rbio 2.8 mother wavelet function.

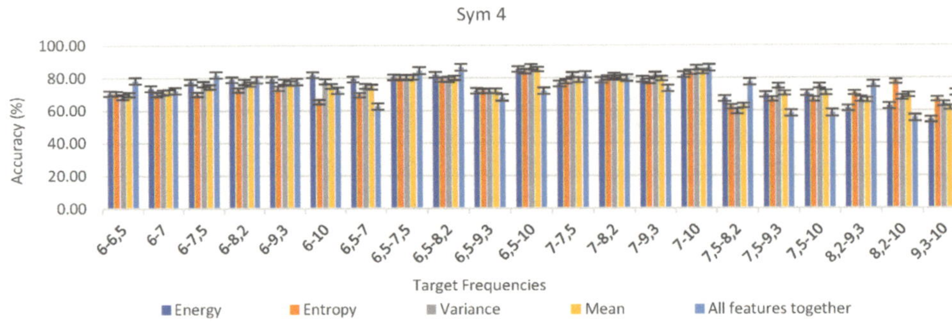

Figure 15.
Binary classification performance of the features for Sym 4 mother wavelet function.

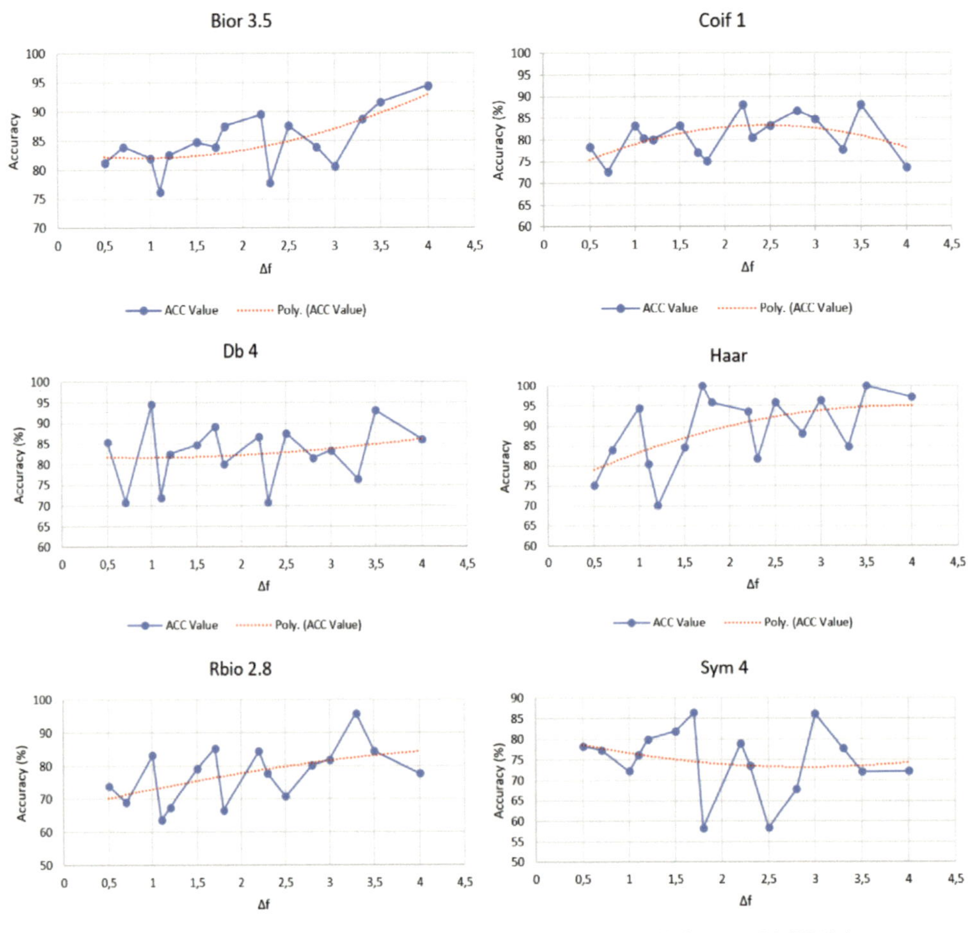

Figure 16.
Change of accuracy value according to the differences between frequencies for mother wavelet functions.

3.1.1 Multiple classification results

Presented in **Table 2** are accuracy results for multiple classification. In regard to these results, the highest performance was shown by the Ensemble Learning classifier with 52.40%.

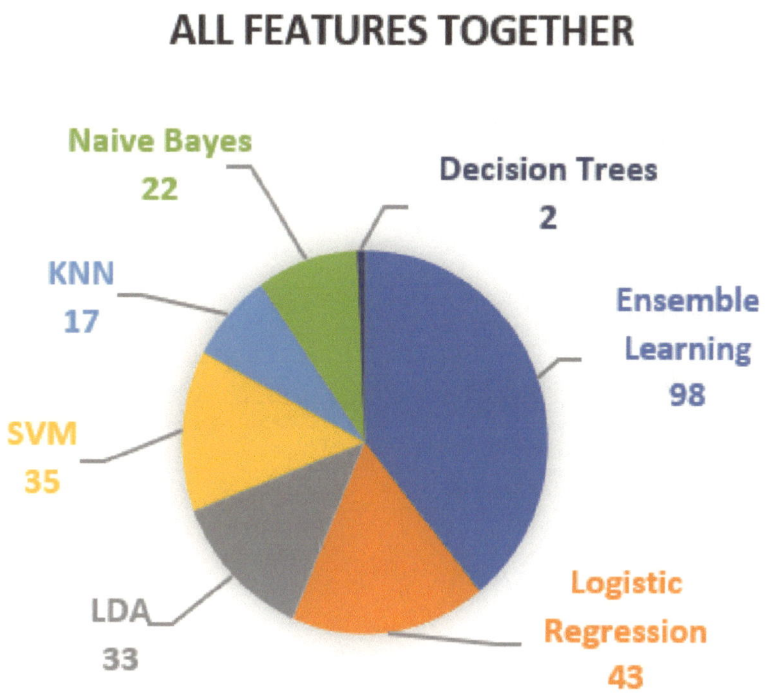

Figure 17.
Percentage of classifier where the best result is the most often obtained as a result of running the algorithms 2,520 times in total (for Haar wavelet function).

Subjects	ACC	Classifier
Subject 1	25.90	LDA
Subject 2	50.00	Ensemble
Subject 3	**52.40**	Ensemble
Subject 4	42.90	Ensemble
Mean	42.80	

Table 2.
Results of multiple classification for time-domain features.

Subjects	ACC	Classifier
Subject 1	29.20	Ensemble
Subject 2	50.00	Ensemble
Subject 3	**57.10**	Ensemble
Subject 4	47.60	Ensemble
Mean	45.98	

Table 3.
Results of multiple classification for frequency-domain features.

3.1.2 Binary classification results

According to the binary classification results shown in **Figure 2**, the best performance was obtained with an accuracy value of 91.68% in 6–10 Hz frequency

	Subject 1		Subject 2		Subject 3		Subject 4	
Mother wavelet	**ACC**	**Classifiers**	**ACC**	**Classifiers**	**ACC**	**Classifiers**	**ACC**	**Classifiers**
Coif 1	29.20	KNN	34.60	Ensemble	33.30	Ensemble	33.30	LDA
Bior 3.5	**55.60**	**LDA**	23.10	Ensemble	42.90	Ensemble	28.60	Naive Bayes
Db 4	37.50	SVM	23.10	SVM	33.30	Naive Bayes	33.30	Ensemble
Sym 4	29.20	LDA	30.80	Decision Tree	38.10	Ensemble	28.60	LDA
Haar	37.50	KNN	23.10	LDA	42.90	LDA	23.80	LDA
Rbio 2.8	33.30	Naive Bayes	23.10	SVM	38.10	Ensemble	28.60	Ensemble
Mean	37.05		26.30		38.10		29.37	

Table 4.
Multiple classification results of wavelet features.

Frequency pair	Energy	Entropy	Variance	Mean	All features together
6–10	95.83	94.45	**100.00**	96.76	97.23
6.5–8.2	92.85	95.83	95.83	94.83	**100.00**
6.5–10	**100.00**	84.50	95.83	93.44	**100.00**

Table 5.
Classification results of the most successful frequency pairs of the Haar mother wavelet.

pairs based on the average of the subjects. Simultaneously, when the subjects are considered separately, a classification performance up to 100% were obtained. In addition, there is no definitive finding related to the increase in the accuracy value parallel to the difference between frequencies for the time-domain.

The results of classifiers to be expressed in the pie chart in **Figure 3** are the number of hits of the classifiers obtained. These numbers were obtained by running all algorithms 2,520 times in total. The best classification performance is shown by the Ensemble learning classifier.

3.2 Frequency-domain features results

For the frequency-domain characteristics used in the problem of determining seven different frequencies, firstly, spectrum analysis was performed to detect the stimulus frequencies more clearly than the signal. This analysis is often used to obtain frequency information in evoked SSVEP responses. The power spectrum of SSVEP signals was determined by FFT using MATLAB software to calculate its power, entropy, and variance for each band in the frequency range corresponding to the frequencies. For this purpose, the signal received FFT is divided into EEG bands (delta, theta, alpha, beta, gamma), and energy, entropy, and variance values of each band are calculated. A total of 15 feature vectors are generated.

3.2.1 Multiple classification results

According to the multiple classification results of the seven frequencies presented in **Table 3**, it was determined that the best performance was in the

Ensemble Learning classifier with an accuracy value of 57.10%. Another remarkable finding here is that the results of the classifier from all individuals are the same. This shows us that, like the time-domain, the Ensemble Learning classifier performs better than others. In addition, when multiple classification results of frequency-domain features are compared with multiple classification results of time-domain features, it has been determined that there is an increase of 4.70% on an individual basis and 3.18% on average.

3.2.2 Selected three class classification results

In this part, three frequencies (6 Hz - 8.2 Hz - 10 Hz), which are considered to increase the classification performance, were chosen among the seven frequencies present in the data set, during the feature extraction phase. The reason for choosing these frequencies are the results of the study done in Ref. [12, 13, 20].

According to the results obtained (**Figure 4**), the highest classification performance for the first participant was 83.30% in the Ensemble Learning classifier, the highest 100% classification performance for the second participant was in the KNN and SVM classifiers, and 88.90% for the third participant in the KNN classifier. Finally, in the fourth participant, it was seen again in the Ensemble Learning classifier with 77.80%.

When the results are evaluated considering the classifiers, the performance of the six different classifiers was calculated by taking the average of the four participants and the highest performance was found in the Ensemble Learning classifier with an accuracy of 79.73%.

3.2.3 Binary classification results

Considering the averages of the binary classification results of frequency features, the performances obtained vary between the lowest 70.85% and the highest 100%. Accordingly, the highest performance was determined with 100% accuracy value in 7.5–10 frequency pairs.

When the results are evaluated in terms of classifiers, it is clearly seen in **Figure 6** that the classifier with the highest accuracy rate is the Ensemble Learning classifier. Runner-up classifier is the SVM classifier. Other classifiers following Ensemble Learning and SVM were identified as KNN, Logistic Regression and Naive Bayes classifiers, in order. It is also seen that no successful results have been obtained in the LDA and Decision Tree classifiers.

3.3 Wavelet transform features results

This section aims to analyze three crucial features, such as energy, variance, and entropy, which are frequently used in DWT studies, have been extracted from the bands (delta, theta, alpha, beta, and gamma) of the EEG signal. These features were generated for six different mother wavelets (Haar, db4, sym4, coif1, bior3.5, rbio2.8) commonly used in the literature. The results of each were evaluated in detail for multiple, binary, and three selected frequencies.

3.3.1 Multiple classification results

On the basis of mother wavelet selection, the results in (**Table 4**), reveal that Bior3.5 and Coif1 mother wavelets were relatively successful, although there is no dominant wavelet type. Experimenting with a larger sample size (number of subjects), in order to generalize, can help obtain more precise results.

In contrast to the mother wavelet selection, when the classifiers are evaluated, the success of Ensemble learning and LDA classifiers is clearly seen.

3.3.2 Classification results for three selected frequencies

In this analysis, as in the classification of frequency-domain features (Section 3.2.2), multiple classification was made by selecting 3 selected frequencies (6 Hz - 8.2 Hz – 10 Hz) where the differences between the frequencies were higher among the seven frequencies. However, unlike the analysis made in the frequency-domain, the selected features are classified and evaluated both they are used together, that is, when energy, variance and entropy features are used as a single feature vector (all features together, and they are used as separate features. Thus, detailed information about the power, irregularity and bias of the signal was obtained. At the same time, it is learned how to use these three features, which have the indispensable properties of the signal, more effectively. And the contribution of these features, which are frequently used in the literature, as a new form of features is wanted to be shown.

In **Figure** 7, the ACC values obtained by classification of the energy, entropy, and variance features extracted using each wavelet family are presented. Mean, minimum and maximum values of the classification results were also shown. According to these results, the values given by the Haar wavelet function for energy, entropy, and variance feature groups, which yield more successful results than other wavelet functions, were 75.85%, 73.08%, and 73.75%, respectively. There were no major differences between the mean values of the features extracted based on the Haar wavelet. However, it was seen that the entropy feature group had a 100% success rate compared to the others.

In **Figure 8**, the extracted features based on wavelet were used as a feature set, and the successful performances of the wavelet families were compared in this way. It was seen that the most successful wavelet family was the Haar wavelet function. The ranking of success in other wavelet families has not changed. The accuracy values are as follows: 75.85% with Haar mother wavelet, 67.53% with bior3.5 mother wavelet, 60.85% with db4 mother wavelet, 56.25% with coif1 mother wavelet, 52.35% with rbio2.8 mother wavelet and 44.73% with sym4 mother wavelet obtained. It was seen that some mother wavelet performances increased when compared with the ACC values in which the features in **Figure** 7 were handled separately. Mean values of coif1, db4, and sym4 mother wavelet functions increased.

As a result of the classification processes performed separately for each subject, when the performances of both feature groups were examined, the most successful wavelet function was found as the Haar wavelet. When the average accuracy values of the feature groups are examined, the results in the case that the three features are used as a single feature vector gave higher results for all wavelet functions than the other feature group. Although there is no dominant result in the comparison of energy, entropy, and variance features among themselves, the highest result was seen in the entropy feature in Subject 3 with 100%.

The results of classifiers to be expressed in the pie chart in **Figure 9** are the number of hits of the classifiers obtained. With reference to results obtained, it is obvious that the most successful and also the most frequent classifier in the classification was obtained as the Ensemble classifier.

3.3.3 Binary classification results

In this analysis, feature vectors are treated as a single feature vector and individual (separate) feature vectors, similar to those in Section 3.2.3. The resulting feature vectors were then evaluated by binary classification in order to analyze frequencies in detail. As the results of the experimental design, the classification performances are obtained for:

- three features separately (energy, entropy and variance),
- average of the three features separately (Mean),
- the extracted features were grouped as a single feature set (All features together).

Each feature (energy, entropy, variance and all features together) extracted using each wavelet family. All values of the classification results are presented in **Figures 10–15** for each mother wavelet, respectively.

According to these results, features obtained from the Haar wavelet function yielded higher accuracies than those obtained from the other wavelet functions. Maximum accuracy performances were obtained in the frequency pairs "6–10", "6.5–8.2", "6.5–10" in the Haar wavelet (**Table 5**). When the features are evaluated, it is realized that the "All features together" feature generally has better results for all mother wavelet functions.

And another researched hypothesis results are presented in **Figure 16** for each mother wavelet, respectively. The purpose here is to show the change in the accuracy value according to the increase in the difference between the frequencies.

Finally, classification results obtained are presented in **Figure 17**. Since the classification results of all the features ranking are similar for all the wavelet functions, the classification result of the "All features together" for Haar wavelet function is presented. According to these results, the most successful classifier was obtained as the Ensemble classifier.

4. Conclusions

This chapter aimed to achieve significant optimization of cortical visual responses, signal processing methods, and ML algorithms, as well as the accuracy and reliability of the superior multi-command SSVEP-based BCI system. New approaches have been explored using existing methods to develop an accurate, reliable, comfortable SSVEP-based BCI that can offer people with severe motor neuron diseases a communication alternative using attention modulation without requiring neuromuscular activities or eye movements.

As a result, the following research objectives were achieved in this study:

- When the results of the time-domain features are evaluated first, it can be seen that these features give usable (noteworthy) results in the classification of SSVEP signals. However, given the natural structure of the SSVEP signal, it is a fact that the results obtained are not sufficient for a real-time SSVEP-based BCI design, since the time-domain properties do not reflect the characteristics of the signal alone.
- According to the classification results of the frequency-domain features, were evaluated alone, satisfactory results were obtained. Higher accuracy values were obtained in both multi-classification and binary classification compared to time-domain.

- And when the last feature group time-frequency domain features are used, using mother DWT functions, SSVEP signals are divided into frequency bands and energy, entropy and variance values of each band are calculated. In this way, feature vectors were created and feature vectors were used as, both separately and also together. Extracted feature vectors were tested with a binary, multiple and three selected classes classification method to see the relationship between seven different classifiers and each frequency in detail. Although multiple classification results seem to be low for all feature groups, there is no study with 7 frequencies (by command) when the literature is searched according to the best knowledge of the author, but high results were obtained compared to studies with 3 and 4 frequencies.

- For stimulation frequency detection in the SSVEP signal, a new form has been proposed that has been proven to be more effective with respect to the use of energy, entropy and variance features than the properties derived from the frequency domain and time-frequency domain. According to this form, instead of the energy, entropy and variance properties used separately, the feature vector, which is all features together, gave better results than the others.

- By conducting detailed research on stimulation frequencies, frequency pairs estimated with the highest accuracy were determined. Although this result showed small differences between the mother wavelet functions, the highest performance was obtained in the frequency pairs in which the difference was generally high (6–10, 6.5–10, 7–10, and 7.5–10 Hz).

- In the literature, the performances of the classifier types that were not compared before were evaluated in terms of SSVEP detection and the most successful classifier was found to be the "Ensemble Classifier".

- Also, does system performance increase in parallel with the differences between frequencies? Based on this hypothesis, the relationship between frequencies was investigated in pairs. A decrease in "Sym4" function was observed, where only the lowest performances were obtained.

- Finally, the most successful mother wavelet selection was made. Accordingly, it was the Haar wavelet function that gave the best results compared to others.

Conflict of interest

The authors declare no conflict of interest.

Thanks

We would like to thank Adnan Vilic for his support in providing SSVEP records.

Author details

Ebru Sayilgan[1*], Yilmaz Kemal Yuce[2] and Yalcin Isler[3]

1 Izmir University of Economics, Izmir, Turkey

2 Alanya Alaaddin Keykubat University, Antalya, Turkey

3 Izmir Katip Celebi University, Izmir, Turkey

*Address all correspondence to: ebru_drms@hotmail.com

References

[1] Wolpaw JR, Boulay CB. Brain signals for brain–computer interfaces. In: Graimann B., Pfurtscheller G., Allison B, editors. Brain-Computer Interfaces. The Frontiers Collection. Springer: Heidelberg; 2009. p. 29-46. DOI: 10.1007/978-3-642-02091-9_2

[2] Graimann B, Allison B, Pfurtscheller G. Brain–computer interfaces: A gentle introduction. In: Graimann B., Pfurtscheller G., Allison B, editors. Brain-Computer Interfaces. The Frontiers Collection. Springer: Heidelberg; 2010. p. 1-27. DOI: 10.1007/978-3-642-02091-9_1

[3] Mason SG, Birch GE. A general framework for brain-computer interface design. IEEE Transactions on Neural Systems and Rehabilitation Engineering. 2003;11(1): 70-85. DOI: 10.1109/TNSRE.2003.810426

[4] Ramadan RA, Vasilakos AV. Brain computer interface: Control signals review. Neurocomputing. 2017;223: 26-44. DOI: 10.1016/j.neucom.2016.10.024

[5] Abiri R, Borhani S, Sellers E, Jiang Y, Zhao X. A comprehensive review of EEG-based brain-computer interface paradigms. Journal of Neural Engineering. 2019;16 011001. DOI: 10.1088/1741-2552/aaf12e

[6] Basar E. EEG-brain dynamics: relation between EEG and brain evoked potentials. 1st ed. Brain Lang Elsevier; 1980. 411 p.

[7] Wang Y, Gao X, Hong B, Jia C, Gao S. Brain-computer interfaces based on visual evoked potentials. IEEE Engineering in Medicine and Biology Magazine. 2008;27(5): 64-71. DOI: 10.1109/MEMB.2008.923958

[8] Regan D. An effect of stimulus colour on average steady-state potentials evoked in man. Nature. 1966;210:1056–1057.

[9] Gao S, Wang Y, Gao X, Hong B. Visual and auditory brain-computer interfaces. IEEE Transactions on Biomedical Engineering. 2014;61(5): 1436–1447. DOI: 10.1109/TBME.2014.2300164

[10] Zhang Y, Xie SO, Wang H, Zhang Z. Data analytics in steady-state visual evoked potential-based brain–computer interface: A review. IEEE Sensors Journal. 2021;21(2):1124-1138. DOI: 10.1109/JSEN.2020.3017491

[11] Huang X, Xu J, Wang Z. A novel instantaneous phase detection approach and its application in SSVEP-based brain-computer interfaces. Sensors. 2018; 18(12):4334. DOI:10.3390/s18124334

[12] Sayilgan E, Yuce YK, Isler Y. Evaluation of wavelet features selected via statistical evidence from steady-state visually-evoked potentials to predict the stimulating frequency. Journal of the Faculty of Engineering and Architecture of Gazi University. 2021;36(2):593-605. DOI:10.17341/gazimmfd.664583

[13] Sayilgan E, Yuce YK, Isler Y. Evaluation of mother wavelets on steady-state visually-evoked potentials for triple-command brain-computer interfaces. Turkish Journal of Electrical Engineering & Computer Sciences. 2021;29(3). DOI:10.3906/elk-2010-26

[14] Sayilgan E, Yuce YK, Isler Y. Investigating the effect of flickering frequency in steady-state visually-evoked potentials on dichotomic brain-computer interfaces. Innovation and Research in BioMedical Engineering. 2021;Under Review.

[15] Zhang Z, Li X, Deng Z. A CWT-based SSVEP classification method for

brain-computer interface system. In: 2010 International Conference on Intelligent Control and Information Processing; 13-15 Aug. 2010; Dalian, China. 2010. pp. 43-48. DOI: 10.1109/ ICICIP.2010.5564336

[16] Bian Y, Li H, Zhao L, Yang G, Geng L. Research on steady state visual evoked potentials based on wavelet packet technology for brain-computer interface. Procedia Engineering. 2011;15: 2629-2633. DOI: 10.1016/j. proeng.2011.08.494

[17] Vilic A. AVI steady-state visual evoked potential (SSVEP) signals dataset 2013 [Internet]. Available from: https://www.setzner.com/avi-ssvep-dataset/. [Accessed 15th August 2018].

[18] Sutter EE. The brain response interface-communication through visually induced electrical brain responses. Journal of Microcomputer Applications. 1992;15(1):31-45.

[19] Bisht A, Srivastava S, Purushothaman G. A new 360° rotating type stimuli for improved SSVEP based brain computer interface. Biomedical Signal Processing and Control. 2020;57: 101778. DOI:10.1016/j.bspc.2019.101778

[20] Sayilgan E, Yuce YK, Isler Y. Prediction of evoking frequency from steady-state visual evoked frequency. Natural and Engineering Sciences. 2019; 4(3): 91-99.

[21] Sayilgan E, Yuce YK, Isler Y. Estimation of three distinct frequencies using fourier transform of steady-state visual-evoked potentials. Duzce University Journal of Science and Technology. 2020;8(4):2337-2343. DOI: 10.29130/dubited.716386

[22] Liu W, Zhang L, Li C. A method for recognizing high-frequency steady-state visual evoked potential based on empirical modal decomposition and canonical correlation analysis. In: 2019 IEEE 3rd Information Technology, Networking, Electronic and Automation Control Conference (ITNEC); 15–17 March 2019; Chengdu, China. 2019. p. 774-778. DOI:10.1109/ ITNEC.2019.8729005

[23] Chen YF, Atal K, Xie SQ, Liu Q. A new multivariate empirical mode decomposition method for improving the performance of SSVEP-based brain-computer interface. Journal of Neural Engineering. 2017;14(4):046028. DOI: 10.1088/1741-2552/aa6a23

[24] Gandhi T, Panigrahi KB, Anand S. A comparative study of wavelet families for EEG signal classification. Neurocomputing. 2011;74(17): 3051-3057. DOI: 10.1016/j. neucom.2011.04.029

[25] Cao Z, et al. Extraction of SSVEPs-based inherent fuzzy entropy using a wearable headband EEG in migraine patients. IEEE Transactions on Fuzzy Systems. 2020;28(1):14-27. DOI: 10.1109/TFUZZ.2019.2905823

[26] Alpaydin E. Introduction to Machine Learning: MIT Press; 2004. 712 p.

[27] Duda RO, Hart PE, Stork DG. Pattern Classification: John Wiley & Sons;2001.

[28] Lotte F, Bougrain L, Cichocki A, Clerc M, Congedo M, Rakotomamonjy A, Yger F. A review of classification algorithms for EEG-based brain-computer interfaces: A 10-year update. Journal of Neural Engineering. 2018;15(3):1-28. DOI: 10.1088/ 1741-2552/aab2f2

[29] Sayilgan E, Yuce YK, Isler Y. Determining gaze information from steady-state visually-evoked potentials. Karaelmas Science and Engineering Journal. 2020;10(2):151-157. DOI: 10.7212/zkufbd.v10i2.1588

[30] Narin A, Isler Y, Ozer M. Comparison of the effects of cross

validation methods on determining performances of classifiers used in diagnosing congestive heart failure. DEÜ Mühendislik Fakültesi Mühendislik Bilimleri Dergisi. 2014;16 (48):1-8.

[31] Jung Y, Hu J. A k-fold averaging cross-validation procedure. Journal of Nonparametric Statistics. 2015;27:1-13. DOI: 10.1080/10485252.2015.1010532

[32] Jiao Y, Du P. Performance measures in evaluating machine learning based bioinformatics predictors for classifications. Quantitative Biology. 2016;4(4): 320–330. DOI: 10.1007/s40484-016-0081-2

[33] Cetin V, Ozekes S, Varol HS. Harmonic analysis of steady-state visual evoked potentials in brain computer interfaces. Biomedical Signal Processing and Control. 2020;60(2020):101999. DOI: 10.1016/j.bspc.2020.101999.

Chapter 9

Brain Computer Interface Drone

Manupati Hari Hara Nithin Reddy

Abstract

Brain-Computer Interface has emerged from dazzling experiments of cognitive scientists and researchers who dig deep into the conscious of the human brain where neuroscience, signal processing, machine learning, physical sciences are blended together and neuroprosthesis, neuro spellers, bionic eyes, prosthetic arms, prosthetic legs are created which made the disabled to walk, a mute to express and talk, a blind to see the beautiful world, a deaf to hear, etc. My main aim is to analyze the frequency domain signal of the brain signals of 5 subjects at their respective mental states using an EEG and show how to control a DJI Tello drone using Insight EEG then present the results and interpretation of band power graph, FFT graph and time-domain signals graph of mental commands during the live control of the drone.

Keywords: Brain Computer Interface, fast Fourier transform, emotiv insight, DJI Tello drone, band power, EEG, Neuroscience, Machine Learning, Signal Processing

1. Introduction

The brain computer interface (BCI) technology makes the possible manipulation of embedded systems using signals generated by brainwaves. A characteristic of the BCI system can easily capture brain signals generated by neural activities, it can also recognize differently firing neural activity patterns, and these signals can transform them into useful commands [1]. These commands can be utilized to control the machines or the devices. BCIs are most commonly applied in prosthetic limbs for paralyzed patients, exoskeletons, robotics, autonomous vehicles, virtual keyboard and computer games [2]. The BCI system can be classified as invasive, non-invasive (these are classified based on the location of placement of EEG biosensors). Non-invasive BCIs are based on electroencephalography (EEG) to record the brain activities using a series of biosensors disposed on the scalp will be able to measure the potential generated by the electrical activity of thousands to billions of cortical neurons inside our brain [3]. Our study is focused on noninvasive BCI using an Electroencephalogram. The neocortex is a convoluted surface which resides at the top of the brain. It is about ⅛ of 1 inch thick. It has 30 billion neurons arranged in 6 layers. Each neuron makes around 10,000 synapses with other neurons, which results in around 300 trillion connections in the total [4]. The most common type of neuron in the cortex is the pyramidal neuron, populations of which are arranged in columns oriented perpendicular to the cortical surface. The surface of the cortex is convoluted, with Fissures sulci, Ridges gyri. The neocortex exhibits functional specialization. Each area of the cortex is specialized for a particular function. The occipital areas near the back of the head specialize in basic visual processing [5]. The parietal areas towards the top of the head specialize in spatial reasoning and motion processing [6]. Visual and auditory recognition occurs in the temporal areas (towards the sides of the head) while frontal areas are involved in planning and higher cognitive functions. Inputs to a cortical area mainly come into the middle layers, Outputs of the cortical area leave from the upper and lower layers [7]. Based on these input–output patterns, the cortex roughly acts as an organized network of sensory, motor areas. Coming to the EEG, it is a device that extracts, organizes, and filters the electric signals which exist due to the neural firings (action potential) of the brain it is used for various diagnosing purposes, it is a popular non-invasive technique for recording the neuronal firing using electrodes placed on the scalp. The currents originating deep in the brain due to the firing of the neurons are not detected by EEG because the voltage fields will fall off with the square of the distance from the source [8]. The time domain signal displays the signals from different electrodes in a graph known as electroencephalography. EEG will reflect the summation of postsynaptic potentials occurring due to firing of thousands of neurons which are oriented radially to the scalp but not due to tangential electrodes. The spatial resolution of EEG is poor in a square centimeter range because of the impedance caused due to the presence of skull, scalp, CSF, meanings [9]. These layers' act as volume conductors and low pass filters to smear the original signals, whereas coming to the temporal resolution is good at the range of milliseconds [9]. This time domain signal from EEG is then converted into frequency domain signal using different transforms in the signal processing such as discrete Fourier transform, fast Fourier transform, etc. The amplified frequencies (according to the fast Fourier transform) which are extracted from brain by electroencephalogram into 4 ranges they are theta (θ) which ranges from 4 Hz to 8 Hz, alpha (α) which ranges from 8 to 12, beta (β) ranges from 12 to 25 Hz and finally gamma (γ) ranges minimum from 25 Hz to maximum of 45 to 75 Hz [10]. After performing many experiments on many patients specifically to observe the type of waves and the amplified frequencies (when will they occur, in what state of patient these waves can be observed) they have presented a generalized form of relation between their frequency ranges and normal human functions. When a person is ready or about to perform tasks or if he/she is in an alert state then more percentage of α frequency waves are generally observed and if a person is task oriented or if he is in a busy state or anxiously thinking or actively concentrating then high percentages of β frequency waves are generally observed, if a person is performing high motor functions, or if the person is switching the activities during multitasking then high percentage of γ frequency waves are observed mostly in the frontal lobe of the human brain. After performing several tests, I was able to predict that in my meditation state a high percentage of θ frequency waves were observed even in a sleeping state where the mind is in a relaxed condition there are high percentages of θ frequency waves. The Emotiv Insight is an EEG Brain wear device which is composed of five sensors that are projected to acquire and measure the key activity from the entire functional areas of the cortex. The device can provide raw EEG Signals, Mental Commands (conscious thoughts), Facial Expressions - Facial mimicry and Measurements of performance of the brain. The principal key characteristics of this scientific design is the dynamic brain-computer interface interactions with more degrees of freedom for controlling physical and virtual objects. The device accurately identifies mental states and emotion such as Engagement, Focus, Excitement, Meditation, Relaxation, Stress [11]. There is a possibility to build brain activity models in real-time based on spatial resolution. A deeper perspective on specific patterns of an individual's brain activity. The very important problem in EEG processing is low signal to the noise ratio since there are many layers between the neural cortex and the scalp and also due to the artifacts which result with great amplitude, the solution to minimize the noise is we can use filtering techniques and noise reduction techniques to remove the noise from the raw EEG data and extract the brain activity signals [12]. Since the EEG signal is non-stationary signal we use classifiers which are trained on user data (which is limited, this is also another main problem) we can generalize those results poorly to the already trained data on the same individual (different for different individuals because of physiological differences, this also limits the use of EEG applications) at different times. The accuracy might increase as we increase the number of training sessions but generalizing for subjects, i.e., to handle inter-subject variability processing pipelines with domain-specific approaches are used in order to clean, extract relevant features and classify (Riemannian geometry based classifiers, adaptive classifiers) EEG data. The subset of Machine Learning which is Deep Learning is used mainly to extract the features, Recently CNNs (convolutional neural networks) are used to simultaneously extract the feature and the classifier in order to achieve end to end supervised feature learning. Hence the devices use CNNs and recurrent neural networks of 3 to 10 layers in total [13].

2. Common EEG patterns

2.1 EEG

In most of the Neurosurgery hospitals or the hospitals where the diagnosis of the human brain takes place, they mostly use EEG which have very high Temporal and spatial resolutions, devices also occupy huge amounts of space. But in the study of Brain Computer Interface wherein we are required to develop the applications to control the external environment in such a way that the patient or subject or the user must not be facing the adaptability issues to significant extent [14]. If we observe the EEG devices which are used in the Hospitals, they generally require minimum of 2 to 3 hours for just equipping the device or placing the electrode in proper locations over the scalp, where in the contact optimizing fluid needs to be applied all over the head and the electrodes needs to be placed which is a tedious and complex process. But recently the company Emotiv has come up with a very sophisticated and easily adaptable device named as INSIGHT. This Insight device is equipped with very efficient specifications, where in which It has 3 axis gyroscope and 3 axis Magnetometer which are very helpful to remove the artifacts due to the head movements which is absent in the case of hospital EEGs (therefore the patients is instructed not to move their head/to avoid the motor movements). There are 5 important electrodes installed in the device which are made of semi-dry polymer. The electrode locations are 2 in the frontal region, 2 in the temporal region and another is at the central peritoneal region. The nomenclature of these electrodes are given according to the international 10–20 system. Where the frontal left electrode is named as AF3, frontal right electrode is named as AF4, temporal left electrode is named as AT7, temporal right electrode is named as AT8 and peritoneal central electrode is named as Pz plus DRL reference mastoid electrode on the left, the channels with built in digital 5th order sinc filter, bandwidth of 0.5-43 Hz with digital notch filters at 50 Hz and 60 Hz, 2.4 Hz wireless connectivity, 8400uV is the dynamic range (input referred), sequential sampling, 128 samples per second is the sampling rate, 14bit motion resolution, the electrodes are semi dry polymers, with 14bits 1 LSB = 0.51muV (16 bit ADC, 2 bits instrumental noise floor discarded) EEG Resolution. Main principle of EEG devices is the differential amplifier which has 2 inputs Input 1 and Input 2. From both the inputs the information is fed, the information can be mathematical or any signal it gives the output resultant signal which is the relative deference chunk of signal. Mounting the device over the subject's scalp is a crucial procedure, where saline glycerol solution is applied to the semi dry electrodes in order to maintain the optimum contact quality with the scalp of the head. The electrodes were being placed in active locations according to the International 10–20 system protocol and the optimal 100% contactivity is ensured. The subject is instructed to not to move to avoid motion artifacts and stay focused on mental commands. Connection between the insight and the laptop is achieved by following the EmotivApp protocol, connection can be established either using Insight dongle or through the Bluetooth connection. Authentic Interpretation of EEG requires a very high amount of training and experience in analyzing and predicting the Graphical data. The most important set of rules which needs to be followed while analyzing the EEG data is the type of montages used, the time domain of the subject state compared to the EEG data at that current time in order to check that there are no external noises or movements made by the subject. There are different types of montages such as longitudinal bipolar Montage, longitudinal-transverse bipolar montage, circumferential bipolar montage, temporal bipolar montage, Cz referential Montage, Ipsilateral ear referential Montage in order to analyze the patient's cognitive state in different ways in order to predict the correct result. In the EmotivePro app in MacBook Air laptop to record raw EEG data for the experiments I have used an Ipsilateral ear referential montage where one input of the differential amplifier will be a DRL reference mastoid electrode and the other input will be any one of the active electrodes. I set up the channel spacing to 400 µV minimum amplitude to −100 µV, maximum amplitude to +100 µV, with a high pass filter.

2.2 EEG pattern in eye blinks

As we know that the cornea is positively charged and the retina is negatively charged, when there is a movement of the eyes towards upwards due to the bell's phenomenon **Figure 1**, when eyes are closed this will result in the abnormal signals in the frontal electrodes [15]. There is an upward peak then downward peak when eye is blinked once, I blinked the eye continuously therefore crist and troughs continuously occurred only at the AF3, AF4 frontal electrodes in the **Figure 2**, this is because of the bells phenomenon when the eyes are closed the cornea moves up due to which there is a change is potential observed near the frontal electrodes, there is an upward peak at the first when the eyes are closed which implies that there is a relative positive potential of AF3/AF4 with the reference electrode and when eyelids opened there is a downward peak which implies that there is a relative negative potential of AF3/AF4 with the reference electrode.

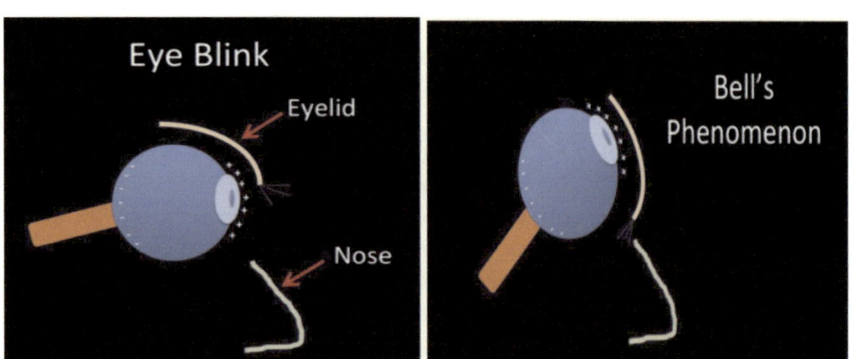

Figure 1.
Bell's phenomenon during eye blink.

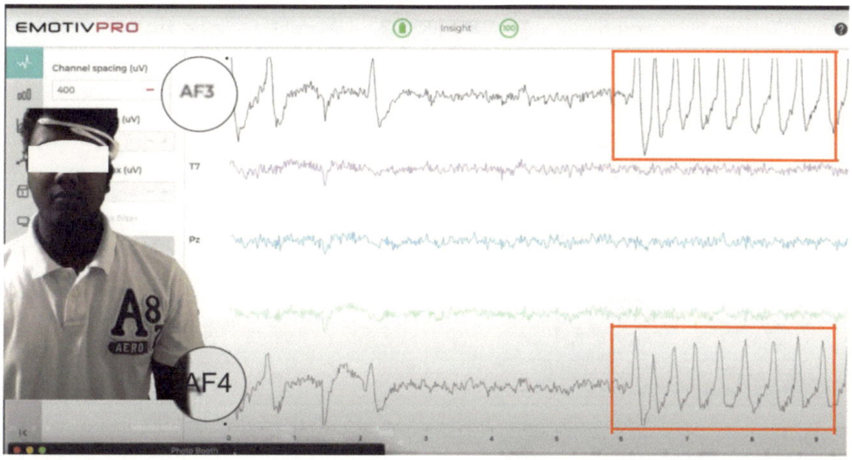

Figure 2.
Crests and troughs observed at AF3,AF4 electrodes in common reference montage EEG data.

2.3 Meditation

To cross check a theoretical aspect of the brain frequencies also in order to validate and prove that the device is authentic, we performed meditation mental tasks for diverse subjects and analyze their signal frequencies of their respective time domain signals using the Fast Fourier Transform (FFT) in the MATLAB.

2.3.1 Test dataset acquisition and observation

The test is being conducted in an anechoic chamber as we can observe at the background of the subjects in Table, in order to minimize the external noise and to isolate the experiment. The subject's voluntary consent has been taken prior for performing this particular test experiment. Also, the subjects are strictly advised to be in a relaxed mental condition for 10 minutes before the test perusal and before 30 seconds of the performance of the test, the subject's eye movements (closed state and opened state) also analyzed by the EmotivPro application in order to remove the eye blink artifacts. All subjects have performed the test successfully as instructed.

As the subject's experiments begin, their mental state is being captured in the camera and the state at which the subjects are present is being noted and presented in the mental state column of **Table 1**. After ensuring the optimum contact quality, the EEG data following common reference montage is analyzed in the live, with EmotivPro license, it is possible to record the current EEG data of its respective electrodes.

After performing the tests on the subjects, the raw EEG data which is recorded during experimentation is stored in the EmotivPro application cloud. This recorded data is exported to the client server system in the .csv format, **Figure 3** these files can be assessed through the links provided in the above **Table 1** for the respective subject test.

This .csv file contains the recorded potentials of all 5 channels (AF3, AF4, T7, T8, Pz) at their respective timestamp. This data is copied and imported into the workspace of MATLAB. Database toolbox is used to read, write, import and export the data of .csv files **Figure 4**. Digital signal processing toolbox is used for converting time domain signal to frequency domain signal.

The main principle of fast Fourier transform is it converts the time domain signal into frequency domain signal. As the raw data is injected into the FFT in matlab, it analyzes the frequencies of the time domain signals. The xfft vs. absolute

SUBJECT Testing Photograph	Subject Name	Mental State	csv file link	Duration
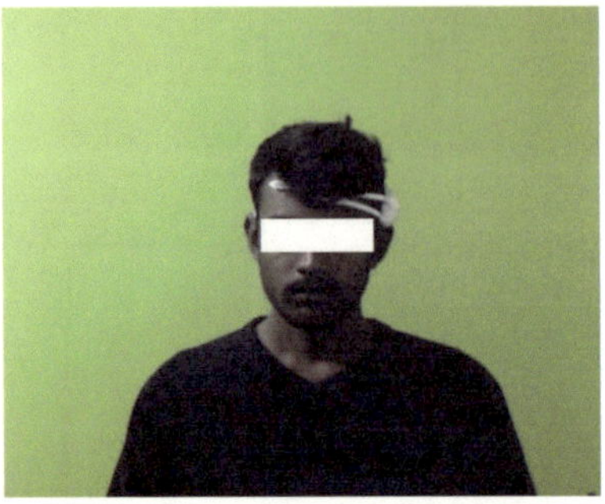	Onesimus	Meditation	OnesimusLink	5 min
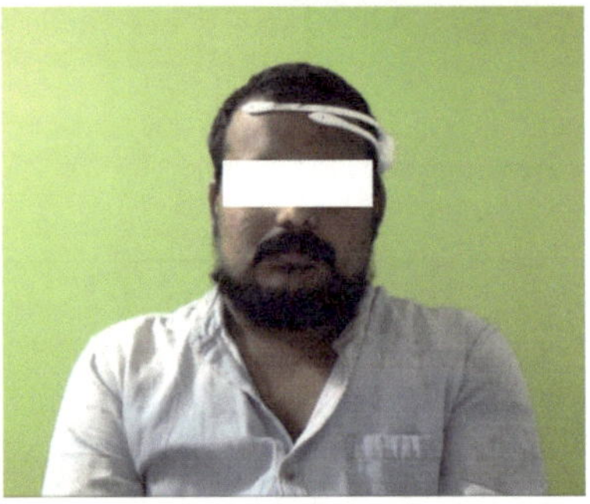	Sai Kumar	Meditation	SaiLink	3 min
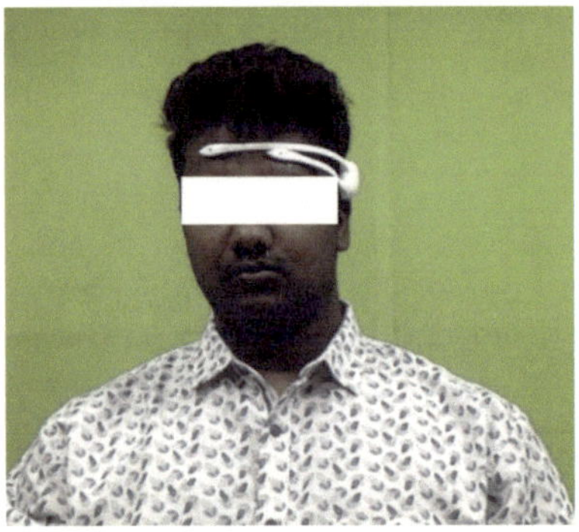	Nithin Reddy	Meditation	NithinLink 1 NithinLink 2 NithinLink 3	3 min
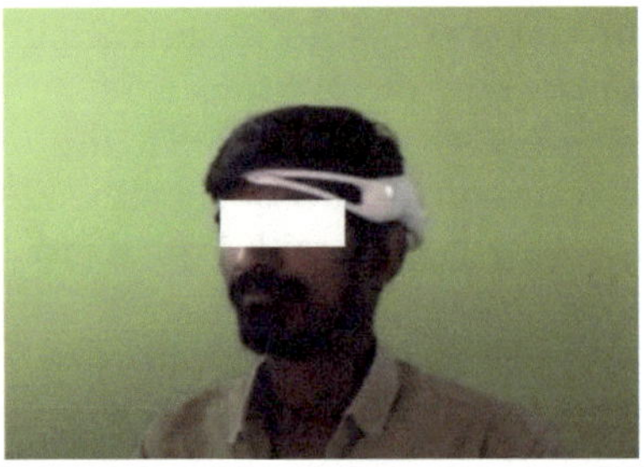	Murty	Relaxed	MurtyLink	5 min

Table 1.
The table displays the specific subject with their respective mental task.

part of the FFT gives us the frequency domain signal. In this case, there is a large amount of noise observed in the frequency domain graph for the test dataset. Hence, in order to remove the noise and the high peak which is near to zero, the smoothing filters were applied independently and a band pass filter is also applied between 4 to 45 Hz, as the result of the experiment lies within that particular frequency range.

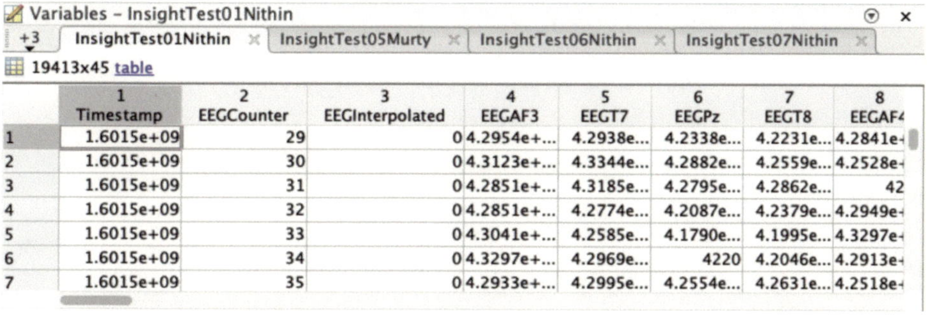
Variables - InsightTest01Nithin
+3 | InsightTest01Nithin | InsightTest05Murty | InsightTest06Nithin | InsightTest07Nithin
19413x45 table

	1 Timestamp	2 EEGCounter	3 EEGInterpolated	4 EEGAF3	5 EEGT7	6 EEGPz	7 EEGT8	8 EEGAF4
1	1.6015e+09	29	0	4.2954e+...	4.2938e...	4.2338e...	4.2231e...	4.2841e
2	1.6015e+09	30	0	4.3123e+...	4.3344e...	4.2882e...	4.2559e...	4.2528e
3	1.6015e+09	31	0	4.2851e+...	4.3185e...	4.2795e...	4.2862e...	42
4	1.6015e+09	32	0	4.2851e+...	4.2774e...	4.2087e...	4.2379e...	4.2949e
5	1.6015e+09	33	0	4.3041e+...	4.2585e...	4.1790e...	4.1995e...	4.3297e
6	1.6015e+09	34	0	4.3297e+...	4.2969e...	4220	4.2046e...	4.2913e
7	1.6015e+09	35	0	4.2933e+...	4.2995e...	4.2554e...	4.2631e...	4.2518e

Figure 3.
The imported data in the workplace.

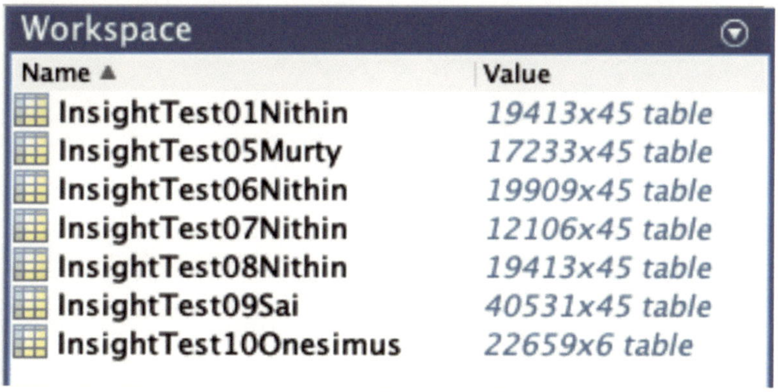
Workspace

Name ▲	Value
InsightTest01Nithin	19413x45 table
InsightTest05Murty	17233x45 table
InsightTest06Nithin	19909x45 table
InsightTest07Nithin	12106x45 table
InsightTest08Nithin	19413x45 table
InsightTest09Sai	40531x45 table
InsightTest10Onesimus	22659x6 table

Figure 4.
The subject's EEG data in the work space.

1. The length of the signal is assigned to variable = nfft,

2. In order to have a good resolution of the signal the nearest 2 power value of the length nfft is assigned to the variable nfft2.

3. The fast Fourier transform filter is applied and assigned to the variable fft.

4. In the absolute part of the fft, we can observe a symmetric curve with frequency x axis and amplitude magnitude of the y-axis.

5. In order to remove the symmetry about the line parallel to the y-axis and only consider the initial half part, we only consider half of the nfft2 length and that half portion is assigned to the variable = fff.

6. For the x component of the fft signal is assigned to the variable xfft, where input values of the sample frequency with half of the length of the data is assigned to xfft.

7. The graphical plot is made with xfft as x-axis and fff as y-axis for that particular time period. Savitzky Golay filter, moving average filter, smoothing filter were applied and analyzed independently to all of the test files. After applying filters to the signals, their respective frequencies are extracted, recorded and saved and the smoothing filter is considered to be the best fit filter at a window of 100 neighboring samples for our test datasets.

<table>
<tr><th>File Name</th><th>Experimental analysis</th></tr>
<tr><td>InsightTestNithin01</td><td>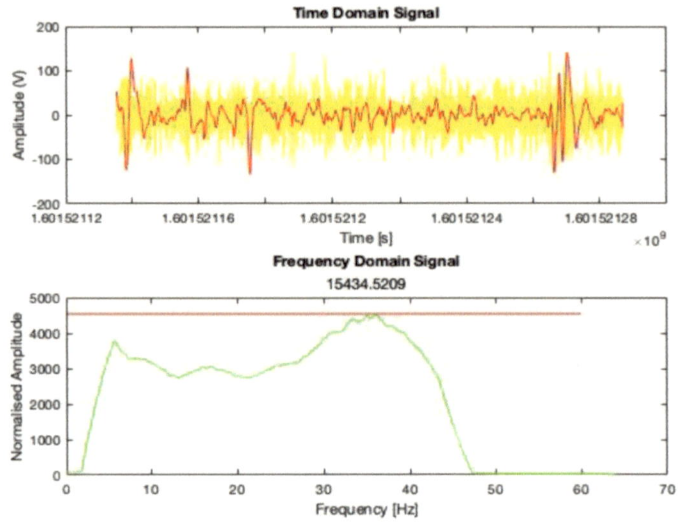
</td></tr>
<tr><td>InsightTestMurty05</td><td>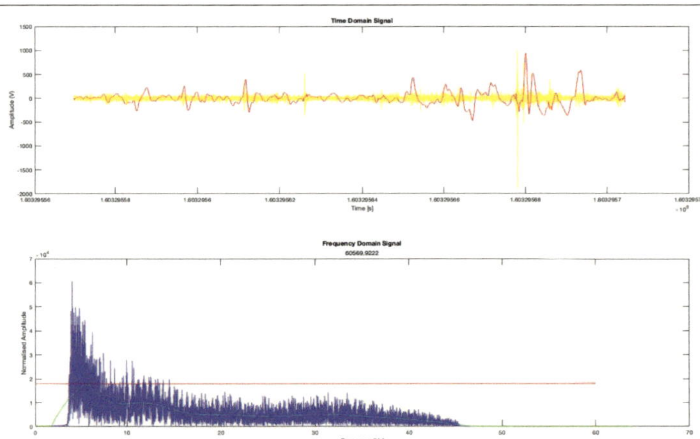</td></tr>
<tr><td>InsightTestNithin06</td><td>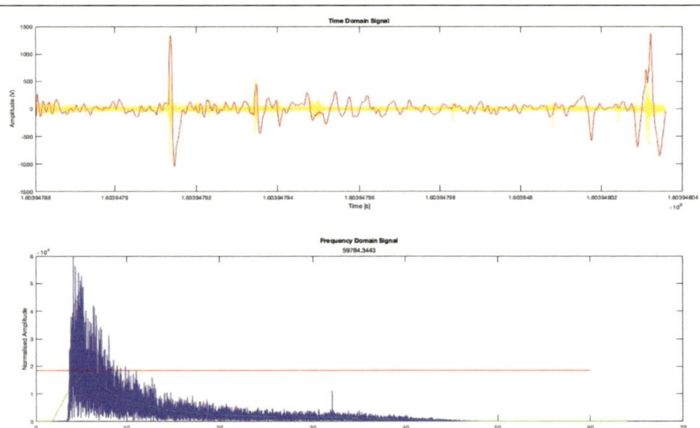</td></tr>
</table>

File Name	Experimental analysis
InsightTestNithin07	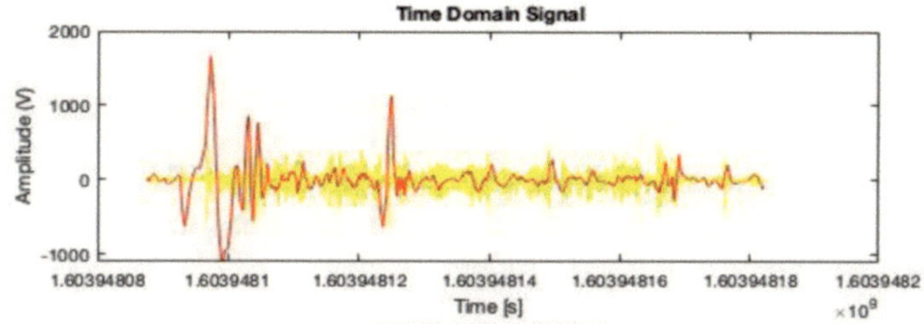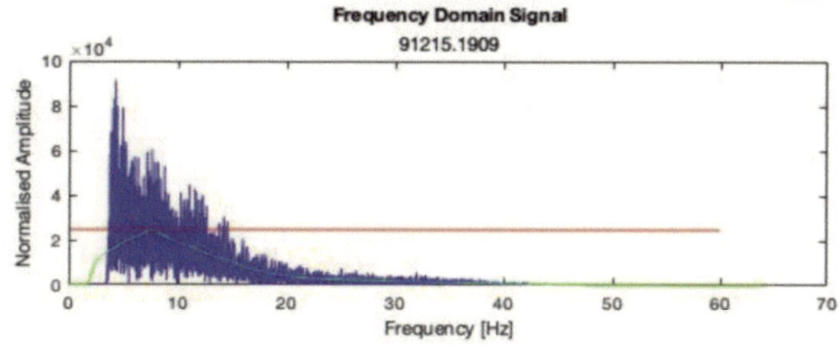
InsightTestNithin08	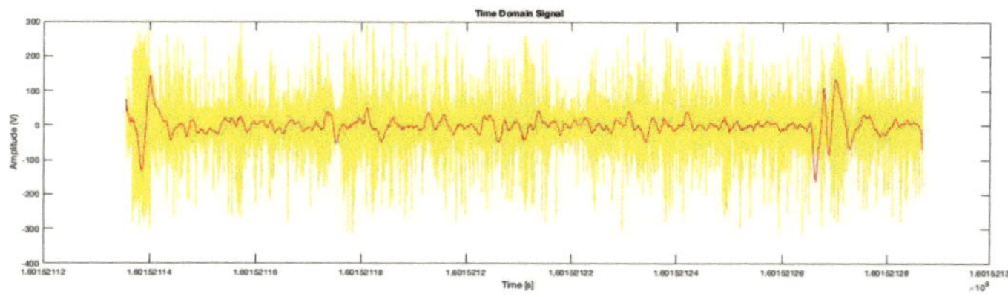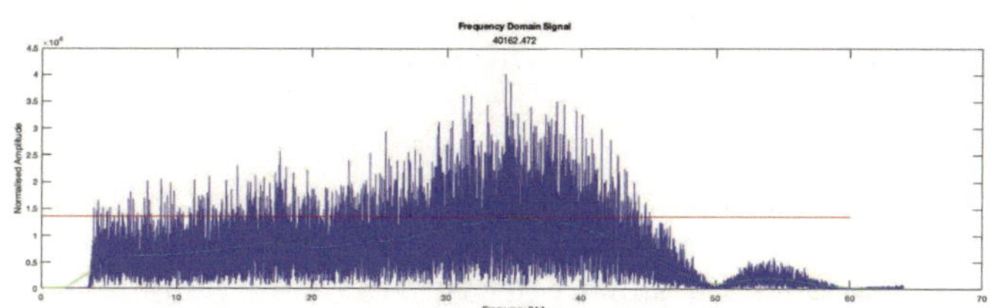
InsightTestSai09	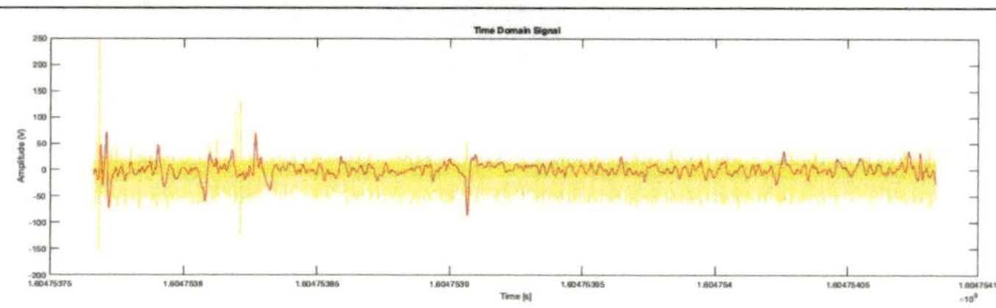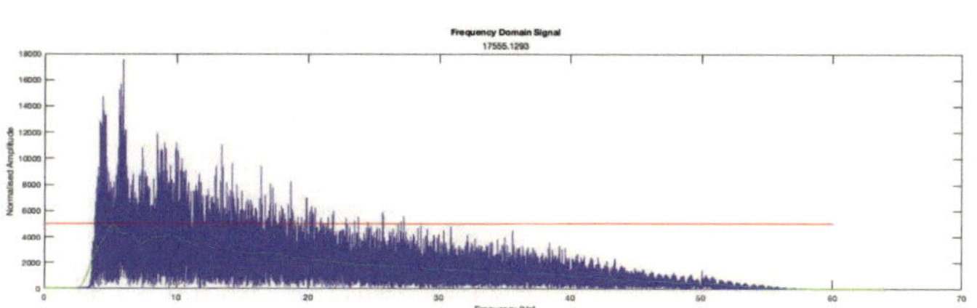

File Name	Experimental analysis
InsightTest Onesimus10	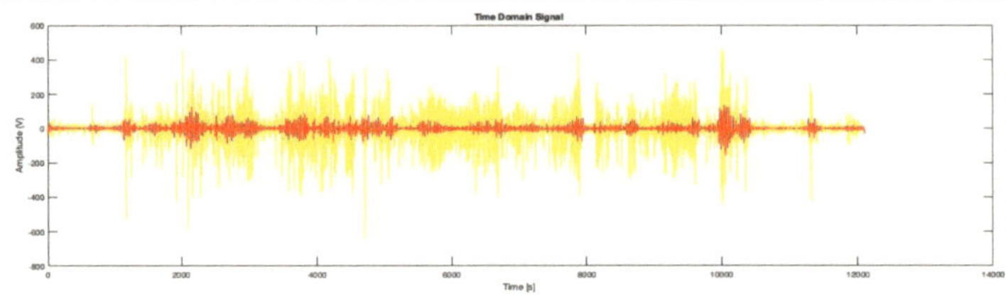 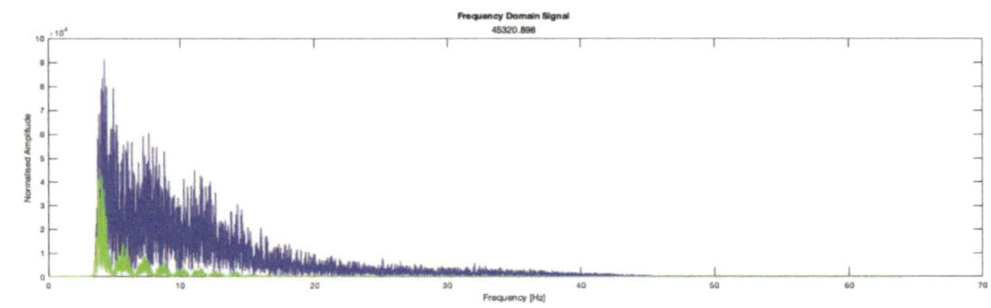

Table 2.
This table represents the experimental analysis graphically.

2.3.2 Test results

The smoothing filter, band pass filter along with FFT in combination resulted in best output for the frequency domain analysis and the results are observed, noted and represented graphically in **Table 2** as there is time domain graph at the top and frequency domain graph at the bottom in the experimental analysis column in **Table 2**.

As most of the subjects are performing the meditative test they should experience a very high amount of theta waves they ranged between 5.1 to 7.6 Hz i.e., there will be a maxima peak within the range of 5.1 to 7.6 Hz in the frequency domain signal and also when the subject performs a demanding motor task he experienced a high amount of gamma waves with maxima peak at 35.9 Hz (**Table 3**).

Hence, the test results prove that the subject's mental state falls within specific range when a specific mental task is performed. This ends the validation part of the Emotiv Insight EEG device.

S. No	File Name	Mental State	Frequency at Maxima (x-axis)	Maxima (y-axis)
1	InsightTestNithin01	Mathematical Calculation	35.9 Hz	45375.0
2	InsightTestMurty05	Meditating	05.5 Hz	17920.0
3	InsightTestNithin06	Meditating	05.6 Hz	18475.0
4	InsightTestNithin07	Meditating	07.6 Hz	24568.0
5	InsightTestNithin08	Lifting	35.9 Hz	13585.0
6	InsightTestSai09	Meditating	05.1 Hz	5016.8
7	InsightTestOnesimus10	Meditating	05.6 Hz	8094.6

Table 3.
This table represents the frequency analysis of subjects numerically.

DOI: http://dx.doi.org/10.5772/intechopen.97558

3. Common EEG patterns

After connecting the device with the EmotivApp, the EmotiveBCI application is used to train mental commands. After training, the live mode is switched ON to control a cube with our Imagery thoughts. Next those live mental commands should be extracted from the application to integrate with the drone. This is achieved through cortex API documentation, for the ease I have edited the code in python language. Then simultaneously those commands should be integrated to a dji tello drone. This is achieved through dji tello API documentation. Finally a link between the drone and Insight brainware is achieved.

3.1 Mental command training

Giving mental command to the EmotivBCI application will result in the movement of the object in the desired direction. Initially numerous training needs to be done to the device in the initial stage to get good desired output. For example during an object movement test when the subject is thinking to move the cube towards the right direction we can observe this in the **Figures 5** and **6** represented below that there is a movement of the cube towards the right direction. Each of the Neutral, moving left, lift, drop mental commands were trained for 10 times.

3.2 Extraction of mental commands and assigning them to the drone

The python code is built in Atom software editor in Dell Inspiron laptop. Jason, websocket, ssl, time, win32, requests, pyautogui, socket, keyboard, threading are the libraries used in the code. The control of the drone with the computer is achieved using DJI Tello drone protocol **Figure 7**, where the drone is connected to the local wifi and the laptop should also be connected to the same wifi using a TP-Link USB Wi-Fi Adapter which is used for PC(TL-WN725N) it's an N150 Wireless Network Adapter for laptop, UDP protocol is used to make an interface between the computer system and dji tello drone, it should be explicit and bind to a local port on our computer where tello can send messages. The functions that listen for messages from tello will be printed on the screen. Connection between the insight and the laptop is done according to the Emotiv App protocol, connection can be established

Figure 5.
The object movement test of the subject, live mode in BCI App when the person is neutral.

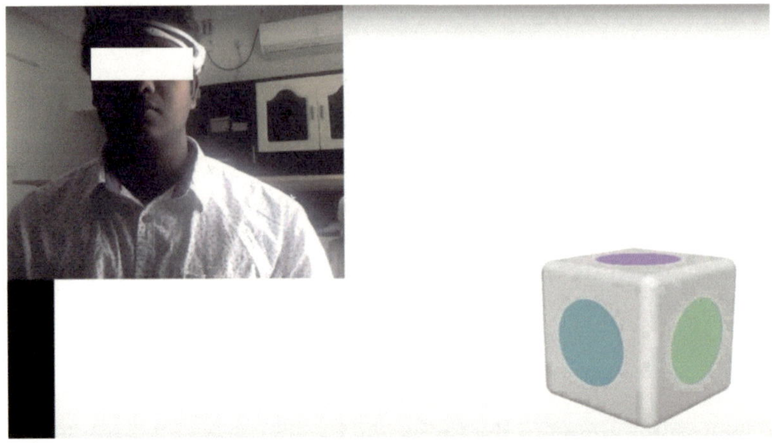

Figure 6.
The object movement test of the subject, live mode in BCI App when the person thinks of moving the object left from neutral, the cube has moved left.

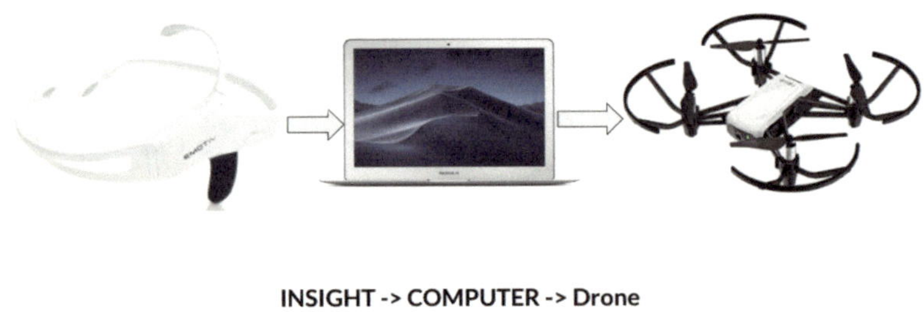

Figure 7.
Procedure of interfacing EEG with a drone.

either using Insight dongle or through the Bluetooth connection. After connecting the device with the EmotivApp, the EmotiveBCI application is used to train mental commands. After training, the live mode is switched ON to control a cube with our Imagery thoughts. Next those live mental commands should be extracted from the application to integrate with the drone. This is achieved through cortex API documentation. Then simultaneously those commands should be integrated to a dji tello drone. This is achieved through dji tello API documentation. Finally a link between the drone and Insight brainware is achieved. Coming to the Insight, InsightHandler class is used then the web connection is acquired, In order to get the approval form the Emotiv app we have to generate client id and secret then we have to give the approval in the emotive app. After approving then authorisation happens to generate the token this token will create the session and loads the profile from the cortex app, then we are going to start and stream the mental commands data in our terminal, these commands will be integrated with the our keyboard clicks, these clicks will intern control the drone as a controller, so basically as we are thinking of a mental command which will be controlling our computers keyboard this will intern control the drone.

4. Result

After several attempts the drone control was modified and whenever the power meter of a mental command crosses 50% then that specific keyboard key will be

DOI: http://dx.doi.org/10.5772/intechopen.97558

clicked. For example if the subject is in a neutral state and suppose he is thinking of a lift command and simultaneously if the power meter of that lift mental command is at 50% or crosses 50% then that command will be executed and the computer keyboard key which is assigned to the push will be clicked this will intern transmits the thrust command from the laptop and drone will receive this command and lifts off the floor. In the terminal window and live mode in the Mental command window controlling a cube are represented in the **Figures 8–10**. The cube is in the

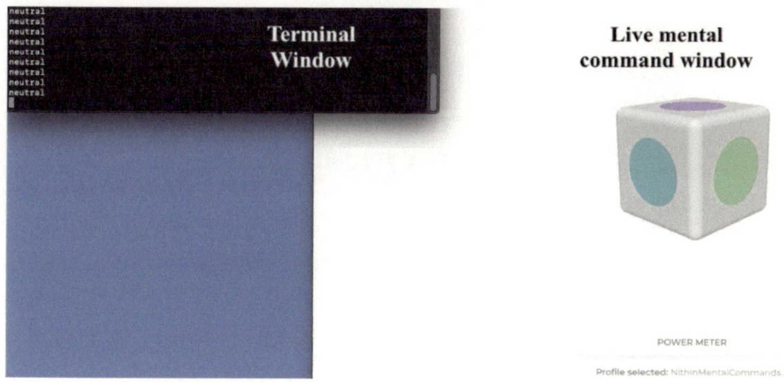

Figure 8.
Cube in neutral state.

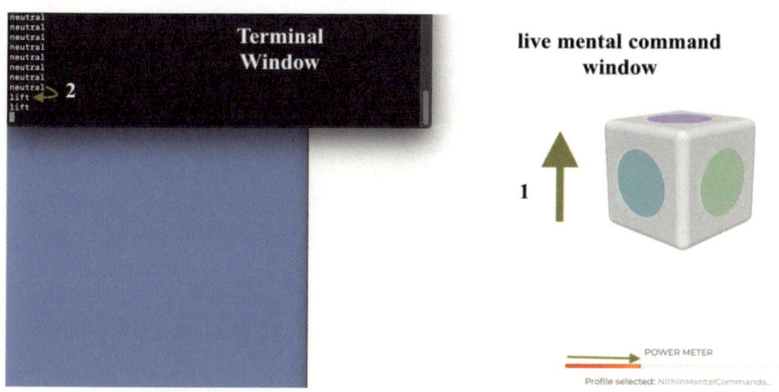

Figure 9.
Cube moving upwards.

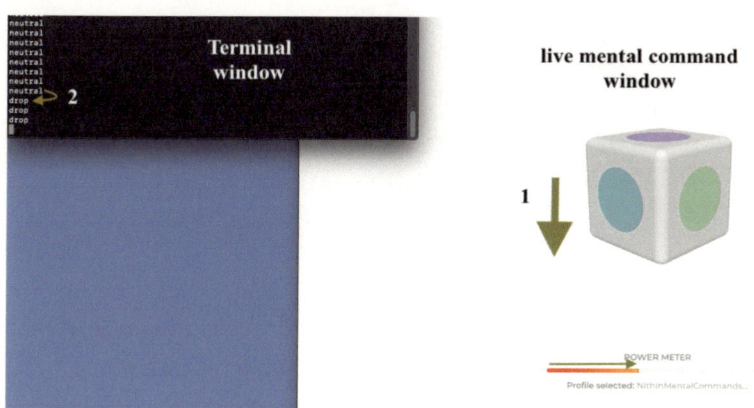

Figure 10.
Cube moving downwards.

Figure 11.
Drone in the neutral state.

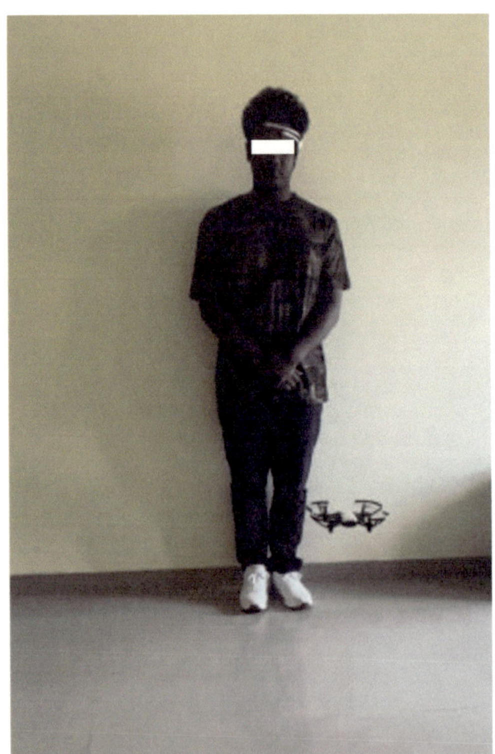

Figure 12.
The drone takes off.

neutral position therefore we can observe neutral in the terminal window in **Figure 8**. When the person is thinking of lifting the cube, the cube changes from its neutral position and moves upwards therefore we can observe change from neutral to lift in the terminal window, **Figure 9**. The cube is in the neutral position therefore we can observe neutral in the terminal window, **Figure 10**.

The drone is in the resting state when the mental command of the subject is neutral, **Figure 11**. The drone takes off when the mental command of the subject has changed from the neutral to lift, **Figure 12**.

5. Conclusions

EEG patterns such as motor movements, eyes movements, meditation, sleeping tests are recorded and analyzed. A mathematical model is developed in the MATLAB using concepts such as signal processing in order to analyze the theoretical data of the brain frequencies at different mental states. The validation of the device by analysis of raw signal has been performed using signal processing methods such as fast Fourier transform simultaneously applying filters to extract the signal of interest by removing noise. These EEG patterns are analyzed on 5 different subjects and cross validated the data with the theoretical brain frequencies data. After these above experiments, Interface between DJI Tello Drone and Emotiv Insight BCI headset was achieved and the Drone was controlled with 2 mental commands moving up and moving down from the neutral state.

Acknowledgements

Inspiration, motivation and presentation have always played a key role in the success of any venture. I express my sincere thanks to "Prof. S. Murugesan" Associate Professor, Department of Pharmacy, Birla Institute of Technology and Science, Pilani being my supervisor and encouraging me to the highest peak and providing me the opportunity to prepare for this research. I feel to acknowledge my indebtedness and deep sense of gratitude to "Dr. Anantha Krishna Chintanpalli" Assistant Professor, Department of Electrical and Electronic Engineering at BITS Pilani for being co-supervisor whose valuable assistance to me in this course has shaped my present work. I am immensely obliged to my friends for their elevating inspiration, encouraging guidance and kind supervision in the completion of my project. Last, but not the least, my parents are also an important inspiration for me. So with due regards, I express my gratitude to them.

Appendices and nomenclature

CPZ	Central Peritoneal Zero
Cz	Central Zero
DBS	Deep Brain Stimulation
DCES	Direct Electrical Cortical Stimulation
ECoG	Electrocorticography
EEG	Electroencephalogram/Electroencephalography
fMRI	Functional Magnetic Resonance Imaging
Hz	Hertz

InsightTestNithin06	Insight: Device Name. TestNithin: Testing the subject name. 06: Test number
IZ	Inion Zero
kHz	KiloHertz
LTD	Long Term Depression
LTP	Long Term Potentiation
MEG	Magnetoencephalography
MU	Memory Unit
Nz	Nasion Zero
OS	Optical Stimulation
Oz	Occipital Zero
POZ	Peritoneal Occipital Zero
Pz	Parietal Zero
STD	Short Term Depression
STDP	Spike timing dependent plasticity
STF	Short Term Facilitation
STP	Short Term Potentiation
TMS	Transcranial Magnetic Stimulation
TUS	Transcranial Ultrasound Stimulation

Author details

Manupati Hari Hara Nithin Reddy
Department of Pharmacy, Birla Institute of Technology and Science, Pilani, Rajasthan, India

*Address all correspondence to: f20171067@pilani.bits-pilani.ac.in

References

[1] Rao, R. (2013). Introduction. In *Brain-Computer Interfacing: An Introduction* (pp. 1-4). Cambridge: Cambridge University Press. DOI: 10.1017/CBO9781139032803.002.

[2] Al-Quraishi MS, Elamvazuthi I, Daud SA, Parasuraman S, Borboni A. EEG-Based Control for Upper and Lower Limb Exoskeletons and Prostheses: A Systematic Review. *Sensors (Basel)*. 2018;18(10):3342. Published 2018 Oct 7. DOI: 10.3390/s18103342.

[3] McFarland DJ, Wolpaw JR. EEG-Based Brain-Computer Interfaces. Curr Opin Biomed Eng. 2017;4:194-200. DOI: 10.1016/j.cobme.2017.11.004.

[4] Wikipedia contributors, 'Neocortex', *Wikipedia, The Free Encyclopedia,* 11 January 2021, 00:52 UTC, <https://en.wikipedia.org/w/index.php?title=Neocortex&oldid=1017132329> [accessed 12 January 2021]

[5] Wikipedia contributors, 'Occipital lobe', *Wikipedia, The Free Encyclopedia,* 25 March 2021, 16:11 UTC, <https://en.wikipedia.org/w/index.php?title=Occipital_lobe&oldid=1014174641> [accessed 16 January 2021]

[6] Ackerman CM, Courtney SM. Spatial relations and spatial locations are dissociated within prefrontal and parietal cortex. J Neurophysiol. 2012; 108(9):2419-2429. DOI: 10.1152/jn.01024.2011

[7] Jawabri KH, Sharma S. Physiology, Cerebral Cortex Functions. [Updated 2021 Feb 14]. In: StatPearls [Internet]. Treasure Island (FL): StatPearls Publishing; 2021 Jan-. Available from: https://www.ncbi.nlm.nih.gov/books/NBK538496/

[8] Buzsáki G, Anastassiou CA, Koch C. The origin of extracellular fields and currents–EEG, ECoG, LFP and spikes. *Nat Rev Neurosci*. 2012;13(6):407-420. Published 2012 May 18. DOI: 10.1038/nrn3241

[9] Burle B, Spieser L, Roger C, Casini L, Hasbroucq T, Vidal F. Spatial and temporal resolutions of EEG: Is it really black and white? A scalp current density view. Int J Psychophysiol. 2015;97(3): 210-220. DOI: 10.1016/j.ijpsycho.2015.05.004.

[10] Hamzei, Nazanin. 2017. "Analysis of Low-Noise EEG in Search of Functional Gamma Band Correlates." Electronic Theses and Dissertations (ETDs) 2008+. T, University of British Columbia. DOI: 10.14288/1.0345615.

[11] Sánchez-Reolid, R.; García, A.S.; Vicente-Querol, M.A.; Fernández-Aguilar, L.; López, M.T.; Fernández-Caballero, A.; González, P. Artificial Neural Networks to Assess Emotional States from Brain-Computer Interface. Electronics **2018**, *7*, 384. DOI: 10.3390/electronics7120384.

[12] Jiang X, Bian GB, Tian Z. Removal of Artifacts from EEG Signals: A Review. *Sensors (Basel)*. 2019;19(5):987. Published 2019 Feb 26. DOI: 10.3390/s19050987.

[13] Alexander Selvikvåg Lundervold, Arvid Lundervold, An overview of deep learning in medical imaging focusing on MRI, Zeitschrift für Medizinische Physik, Volume 29, Issue 2, 2019, Pages 102-127, ISSN 0939-3889, DOI: 10.1016/j.zemedi.2018.11.002.

[14] Shih JJ, Krusienski DJ, Wolpaw JR. Brain-computer interfaces in medicine. Mayo Clin Proc. 2012;87(3):268-279. DOI: 10.1016/j.mayocp.2011.12.008.

[15] Wikipedia contributors, 'Bell's phenomenon', *Wikipedia, The Free Encyclopedia,* 1 January 2021, 00:32 UTC, <https://en.wikipedia.org/w/index.php?title=Bell%27s_phenomenon&oldid=997534506> [accessed 12 January 2021].